183 Topics in Current Chemistry

Springer-Verlag Berlin Heidelberg GmbH

Density Functional Theory IV

Theory of Chemical Reactivity

Volume Editor: R. F. Nalewajski

With contributions by
M. H. Cohen, K. N. Houk, J. Korchowiec,
A. Michalak, R. F. Nalewajski, O. Wiest

With 46 Figures and 10 Tables

Springer

This series presents critical reviews of the present position and future trends in modern chemical research. It is addressed to all research and industrial chemists who wish to keep abreast of advances in the topics covered.

As a rule, contributions are specially commissioned. The editors and publishers will, however, always be pleased to receive suggestions and supplementary information. Papers are accepted for "Topics in Current Chemistry" in English.

In references Topics in Current Chemistry is abbreviated Top.Curr.Chem. and is cited as a journal.

Springer WWW home page: http://www.springer.de

ISSN 0340-1022

ISBN 978-3-662-14841-9   ISBN 978-3-540-49951-0 (eBook)
DOI 10.1007/978-3-540-49951-0

Library of Congress Catalog Card Number 74-644622

© Springer-Verlag Berlin Heidelberg 1996
Originally published by Springer-Verlag Berlin Heidelberg New York in 1996
Softcover reprint of the hardcover 1st edition 1996

The use of general descriptive names, registered names, trademarks, etc. in this publication does not imply, even in the absence of a specific statement, that such names are exempt from the relevant protective laws and regulations and therefore free for general use.

Typesetting: Macmillan India Ltd., Bangalore-25
SPIN: 10537017   66/3020 – 5 4 3 2 1 0 – Printed on acid-free paper

Foreword

Density functional theory (DFT) is an entrancing subject. It is entrancing to chemists and physicists alike, and it is entrancing for those who like to work on mathematical physical aspects of problems, for those who relish computing observable properties from theory and for those who most enjoy developing correct qualitative descriptions of phenomena in the service of the broader scientific community.

DFT brings all these people together, and DFT needs all of these people, because it is an immature subject, with much research yet to be done. And yet, it has already proved itself to be highly useful both for the calculation of molecular electronic ground states and for the qualitative description of molecular behavior. It is already competitive with the best conventional methods, and it is particularly promising in the applications of quantum chemistry to problems in molecular biology which are just now beginning. This is in spite of the lack of complete development of DFT itself. In the basic researches in DFT that must go on, there are a multitude of problems to be solved, and several different points of view to find full expression.

Thousands of papers on DFT have been published, but most of them will become out of date in the future. Even collections of works such as those in the present volumes, presentations by masters, will soon be of mainly historic interest. Such collections are all the more important, however, when a subject is changing so fast as DFT is. Ative workers need the discipline imposed on them by being exposed to the works of each other. New workers can lean heavily on these sources to learn the different viewpoints and the new discoveries. They help allay the difficulties associated with the fact that the literature is in both physics journals and chemistry journals. [For the first two-thirds of my own scientific career, for example, I felt confident that I would miss nothing important if I very closely followed the Journal of Chemical Physics. Most physicists, I would guess, never felt the need to consult JCP. What inorganic or organic chemist in the old days took the time to browse in the physics journals?] The literature of DFT is half-divided, and DFT applications are ramping into chemical and physical journals, pure and applied. Watch JCP, Physical Review A and Physical Review B, and watch even Physical Review Letters, if you are a chemist interested in applying DFT. Or ponder the edited volumes, including the present two. Then you will not be surprised by the next round of improvements in DFT methods. Improvements are coming.

The applications of quantum mechanics to molecular electronic structure may be regarded as beginning with Pauling's Nature of the Chemical Bond, simple molecular orbital ideas, and the Huckel and Extended Huckel Methods. The molecular orbital method then was systematically quantified in the Hartree-Fock SCF Method; at about the same time, its appropriateness for chemical description reached its most elegant manifestation in the analysis by Charles Coulson of the Huckel method. Chemists interested in structure learned and taught the nature of the Hartree-Fock orbital description and the importance of electron correlation in it. The Hartree-Fock single determinant is only an approximation. Configurations must be mixed to achieve high accuracy. Finally, sophisticated computational programs were developed by the professional theoreticians that enabled one to compute anything. Some good methods involve empirical elements, some do not, but the road ahead to higher and higher accuracy seemed clear: Hartree-Fock plus correction for electron correlation. Simple concepts in the everyday language of non theoretical chemists can be analyzed (and of course have been much analyzed) in this context.

Then, however, something new came along, density functional theory. This is, of course, what the present volumes are about. DFT involves a profound change in the theory. We do not have merely a new computational gimmick that improves accuracy of calculation. We have rather a big shift of emphasis. The basic variable is the electron density, not the many-body wavefunction. The single determinant of interest is the single determinant that is the exact wavefuntion for a noninteracting (electron-electron repulsion-less) system corresponding to our particular system of interest, and has the same electron density as our system of interest. This single determinant, called the Kohn-Sham single determinant, replaces the Hartree-Fock determinant as the wavefunction of paramount interest, with electron correlation now playing a lesser role than before. It affects the potential which occurs in the equation which determines the Kohn-Sahm orbitals, bus once that potential is determined, there is no configurational mixing or the like required to determine the accurate electron density and the accurate total electronic energy. Hartree-Fock orbitals and Kohn-Sham orbitals are quantitatively very similar, it has turned out. Of the two determinants, the one of Kohn-Sham orbitals is mathematically more simple than the one of Hartree-Fock orbitals. Thus, each KS orbital has its own characteristic asymptotic decay; HF orbitals all share in the same asymptotic decay. The highest KS eigenvalue is the exact first ionization potential; the highest HF eigenvalue is an approximation to the first ionization potential. The KS effective potential is a local multiplicative potential; the HF potential is nonlocal and nonmultiplicative. And so on. When at the Krakow meeting I mentioned to a physicist that I thought that chemists and physicists all should be urged to adopt the KS determinant as the basic descriptor for electronic structure, he quickly replied that the physicists had already done so. So, I now offer that suggestion to the chemistry community.

On the conceptual side, the powers of DFT have been shown to be considerable. Without going into detail, I mention only that the Coulson work referred to above anticipated in large part the formal manner in which DFT describes molecular changes, and that the ideas of electronegativity and hardness fall into place, as do Ralph Pearson's HSAB and Maximum Hardness Principles.

It was Mel Levy, I think who first called density functional theory a charming subject. Charming it certainly is to me. Charming it should be revealed to you as you read the diverse papers in these volumes.

Chapel Hill, 1996   Robert G. Parr

Foreword

Thirty years after Hohenberg and myself realized the simple but important fact that the theory of electronic structure of matter can be rigorously based on the electronic density distribution $n(r)$ a most lively conference was convened by Professor R. Nalewajski and his colleagues at the Jagiellonian University in Poland's historic capital city, Krakow. The present series of volumes is an outgrowth of this conference.

Significantly, attendees were about equally divided between theoretical physicists and chemists. Ten years earlier such a meeting would not have had much response from the chemical community, most of whom, I believe, deep down still felt that density functional theory (DFT) was a kind of mirage. Firmly rooted in a tradition based on Hartree Fock wavefunctions and their refinements, many regarded the notion that the many electron function, $\Psi(r_1 \ldots r_N)$ could, so to speak, be traded in for the density $n(r)$, as some kind of not very serious slight-of-hand. However, by the time of this meeting, an attitudinal transformation had taken place and both chemists and physicists, while clearly reflecting their different upbringings, had picked up DFT as both a fruitful viewpoint and a practical method of calculation, and had done all kinds of wonderful things with it.

When I was a young man, Eugene Wigner once said to me that understanding in science requires understanding from *several different points of view.* DFT brings such a new point of view to the table, to wit that, in the ground state of a chemical or physical system, the electrons may be regarded as a *fluid* which is fully characterized by its density distribution, $n(r)$. I would like to think that this viewpoint has enriched the theory of electronic structure, including (via potential energy surfaces) molecular structure; the chemical bond; nuclear vibrations; and chemical reactions.

The original emphasis on electronic ground states of non-magnetic systems has evolved in many different directions, such as thermal ensembles, magnetic systems, time-dependent phenomena, excited states, and superconductivity. While the abstract underpinning is exact, implementation is necessarily approximate. As this conference clearly demonstrated, the field is vigorously evolving in many directions: rigorous sum rules and scaling laws; better understanding and description of correlation effects; better understanding of chemical principles and phenomena in terms of $n(r)$; application to systems consisting of thousands of atoms; long range polarization energies; excited states.

Here is my personal wish list for the next decade: (1) An improvement of the accurary of the exchange-correlation energy $E_{xc}[n(r)]$ by a factor of 3-5. (2) A practical, systematic scheme which, starting from the popular local density approach, can – with sufficient effort – yield electronic energies with any specified accuracy. (3) A sound DFT of excited states with an accuracy and practicality comparable to present DFT for ground states. (4) A practical scheme for calculating electronic properties of systems of 10^3 - 10^5 atoms with "chemical accuracy". The great progress of the last several years made by many individuals, as mirrored in these volumes, makes me an optimist.

Santa Barbara, 1996   Walter Kohn

Preface

Density functional methods emerged in the early days of quantum mechanics; however, the foundations of the modern density functional theory (DFT) were established in the mid 1960 with the classical papers by Hohenberg and Kohn (1964) and Kohn and Sham (1965). Since then impressive progress in extending both the theory formalism and basic principles, as well as in developing the DFT computer software has been reported. At the same time, a substantial insight into the theory structure and a deeper understanding of reasons for its successes and limitations has been reached. The recent advances, including new approaches to the classical Kohn-Sham problem and constructions of more reliable functionals, have made the ground-state DFT investigations feasible also for very large molecular and solid-state systems (of the order of 10^3 atoms), for which conventional CI calculations of comparable accuracy are still prohibitively expensive. The DFT is not free from difficulties and controversies but these are typical in a case of a healthy, robust discipline, still in a stage of fast development. The growing number of monographs devoted to this novel treatment of the quantum mechanical many body problem is an additional measure of its vigor, good health and the growing interest it has attracted.

In addition to a traditional, solid-state domain of appplications, the density functional approach also has great appeal to chemists due to both computational and conceptual reasons. The theory has already become an important tool within quantum chemistry, with the modern density functionals allowing one to tackle problems involving large molecular systems of great interest to experimental chemists. This great computational potential of DFT is matched by its already demonstrated capacity to both rationalize and quantify basic classical ideas and rules of chemistry, e.g., the electronegativity and hardness/softness characteristics of the molecular electron distribution, bringing about a deeper understanding of the nature of the chemical bond and various reactivity preferences. The DFT description also effects progress in the theory of chemical reactivity and catalysis, by offering a "thermodynamic-like" perspective on the electron cloud reorganization due to the reactant/catalyst presence at various intermediate stages of a reaction, e.g. allowing one to examine the relative importance of the polarization and charge transfer components in the resultant reaction mechanism, to study the influence of the infinite surface reminder of cluster models of heterogeneous catalytic systems, etc.

The 30th anniversary of the modern DFT was celebrated in June 1994 in Cracow, where about two hundred scientists gathered at the ancient Jagiellonian University Robert G. Parr were the honorary chairmen of the conference. Most of the reviewers of these four volumes include the plenary lecturers of this symposium; other leading contributors to the field, physicists and chemists, were also invited to take part in this DFT survey. The fifteen chapters of this DFT series cover both the basic theory (Parts I, II, and the first article of Part III), applications to atoms, molecules and clusters (Part III), as well as the chemical reactions and the DFT rooted theory of chemical reactivity (Part IV). This arrangement has emerged as a compromise between the volume size limitations and the requirements of the maximum thematic unity of each part.

In these four DFT volumes of the *Topics in Current Chemistry* series, a real effort has been made to combine the authoritative reviews by both chemists and physicists, to keep in touch with a wider spectrum of current developments. The Editor deeply appreciates a fruitful collaboration with Dr. R. Stumpe, Dr. M. Hertel and Ms B. Kollmar-Thoni of the Springer-Verlag Heidelberg Office, and the very considerable labour of the Authors in preparing these interesting and informative monographic chapters.

Cracow, 1996   Roman F. Nalewajski

Table of Contents

Table of Contents of Volume 180

Table of Contents of Volume 181

Density Functional Theory II: Relativistic and Time-Dependent Extensions

Table of Contents of Volume 182

Density Functional Theory III: Interpretation, Atoms, Molecules and Clusters

Density Functional Theory Calculations
of Pericyclic Reaction Transition Structures

Olaf Wiest[1] and K. N. Houk[2]

[1]Department of Chemistry and Biochemistry, University of Notre Dame, Notre Dame, IN 46556, USA
[2]Department of Chemistry and Biochemistry, University of California, Los Angeles, CA 90095, USA

Table of Contents

Topics in Current Chemistry, Vol. 183
© Springer-Verlag Berlin Heidelberg 1996

1 Introduction

During the last 30 years, the study of the mechanism of pericyclic reactions has been one of the major success stories of theoretical organic chemistry. The use of qualitative MO theory such as the analysis of orbital symmetry by Woodward and Hoffmann [1] and Fukui's frontier molecular orbital concept [2–5] greatly increased our understanding of these important reactions. With the development of powerful computers and the availability of software for quantum chemistry during the last two decades, quantitative MO calculations of larger organic molecules became routinely possible. By using ab initio theory, a detailed picture of the structures and properties of the transition structures of a variety of pericyclic reactions has been obtained [6]. Theory has been actively employed in lively arguments about the mechanisms of these reactions [7].

The majority of the calculations published so far were performed according to Hartree-Fock theory, which neglects electron correlation. Because of this, the calculated activation energies for pericyclic reactions are systematically too high and there could be systematic errors in geometries. Furthermore, the comparison between a concerted reaction and a possible stepwise mechanism involving a diradicaloid intermediate is generally not possible, because RHF theory cannot be used for diradicals, and the energies of the UHF wavefunctions of the diradicaloid species are too low compared to RHF calculations on closed-shell species. The inclusion of electron correlation by Moeller-Plesset (MPn) perturbation theory or truncated configuration interaction (CI) calculations can solve this problem, but such calculations limit the size of the molecule to be studied practically to about 10 nonhydrogen atoms with current computers.

Recently, density functional theory (DFT) has evolved into a viable alternative approach for molecular structure calculations. As has been pointed out elsewhere in this volume, DFT-based methods include electron correlation explicitly in the exchange-correlation functional at a moderate computational cost. These methods are therefore promising for highly accurate calculations of chemically interesting systems. The application of DFT to pericyclic reactions is interesting for several reasons. First, the inclusion of electron correlation in DFT can yield accurate results for activation energies and geometries of reactants and transition structures. Second, the high computational efficiency of these methods and the favorable scaling factors allow the study of larger, more interesting systems to practicing chemists. Thirdly, comparison of the results from DFT calculations with the extensive results from Hartree-Fock calculations and other MO-based methods allows an assessment of the accuracy of the different methods available.

In the present chapter, we review the pericyclic reactions studied with DFT methods to date. Local, nonlocal, and hybrid DFT methods have been used to study the parent systems of the most important pericyclic reactions. These results are compared with results of Hartree-Fock theory, post-Hartree-Fock calculations, and available experimental data. Our aim is to provide an overview of the applicability of DFT methods for the study of this important class of reactions.

2 Electrocyclic Reactions

2.1 The Ring Opening of Cyclobutene

The conrotatory electrocyclic ring opening of cyclobutene to form *trans*-1,3-butadiene (Fig. 1) has been studied extensively using a variety of MO [8–11] and DFT methods [12–16]. It has been pointed out [6] that the geometries obtained by the different MO methods are very similar. All methods predict a C_2 symmetric transition structure with various degrees of bond breaking. The carbon skeleton in the transition structure is considerably twisted out of plane, leading to a diradicaloid character of the carbon termini. The computed lengths for the breaking bond range from 2.13 Å (RHF/6-31G*) to 2.24 Å (two configuration SCF). However, in order to reproduce the experimental values for the activation energy [17–19] and heat of reaction [20, 21], correlated methods and large basis sets are necessary. The RHF/6-31G* method gives an activation energy which is 12 kcal/mol higher than the experimental value.

Table 1 summarizes the results for the calculated transition structure geometries and the reaction energetics using the local, nonlocal, and hybrid DFT methods as well as selected results from MO-based calculations. The local spin density approximation (LDA) calculations [22] using different basis sets [23] yield significantly different values for the length of the breaking bond in the transition structure, while the other geometric parameters are in good agreement with each other and with the values obtained from MO calculations. The activation energies obtained from the LDA calculations are in excellent agreement with the experimental value of 32.9 kcal/mol. The exothermicities for the formation of 1,3-butadiene calculated by LDA methods are, however, too low by 3–5 kcal/mol.

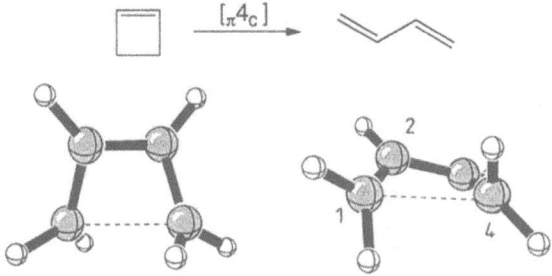

Fig. 1. Transition structure for the electrocyclic ring opening of cyclobutene to *trans*-1,3-butadiene

Table 1. Transition structure geometries, activation energies, and reaction energies for transition structures of electrocyclic ring opening of cyclobutene

Method	Reference	R_{C1-C4} (Å)	R_{C1-C2} (Å)	R_{C2-C3} (Å)	$\phi_{C1-C2-C3-C4}$ (degrees)	E_a (kcal/mol)	ΔE_{react} (kcal/mol)
LDA/DZVP	16	2.142	1.421	1.382	—	33.0	− 6.3
LDA/TZVP	12	2.203	1.417	1.371	20.5	32.0	− 8.3
BP/TZVP	12	2.148	1.433	1.374	20.2	29.7	−14.5
BLYP/6-31G*	13,14	2.161	1.440	1.385	20.0	29.8	−12.5
BP/DZVP	16	2.155	1.437	1.388	—	31.9	−10.7
ACM/6-31G*	13	2.142	1.420	1.379	21.3	36.1	−18.5
Becke3LYP/6-31G*	15	2.144		1.376	20.7	33.9	−12.7
RHF/6-31G*	9,13	2.130	1.413	1.369	22.0	45.2	−10.6
MP4/6-311G**//MP2/6-311G**	12	2.131	1.429	1.384	22.2	33.7	−11.0
CISD/DZVP//TCSCF/DZVP	8	2.238	1.462	1.351	16.4	35.8	
Experiment	17,18	—	—	—	—	32.9 ± 0.5	−11.4 ± 0.5

E_a, activation energy; ΔE_{react}, reaction energy.

The results from the nonlocal and hybrid DFT calculations show trends similar to those observed for the MO calculations. The computed geometries are very similar to each other; the calculated bond lengths for the breaking bond range from 2.14 Å, calculated by the hybrid adiabatic connection method (ACM) [24], to 2.16 Å, calculated by the nonlocal BLYP [25, 26] functional. A comparison of the results from the nonlocal Becke-Perdew [27] calculations using DZVP and a TZVP basis sets shows that the geometries are influenced to a very small extent by basis set effects [12]. The activation energies calculated by the different nonlocal methods are too low by 1–3 kcal/mol, whereas the hybrid DFT methods overestimate E_a by approximately the same amount. Both the nonlocal and the hybrid DFT methods tend to overestimate the heat of reaction by up to 7 kcal/mol, calculated by the ACM/6-31G* method.

The electrocyclic ring opening of cyclobutene was also the first system to be studied by intrinsic reaction coordinate following at the LDA level [28]. Deng and Ziegler calculated the IRC connecting the C_2 symmetric transition structure to cyclobutene and *gauche*-1,3-butadiene [12]. The results from these calculations are shown in Fig. 2a–c. The reaction can be completely described by the simultaneous and gradual variation of four internal modes: The breaking of the σ bond (R_{C1-C4}), the reorganization of the π framework (R_{C1-C2} and R_{C2-C3}), the skewing of the carbon skeleton ($\phi_{C1-C2-C3-C4}$), and the rotation of the methylene groups ($\phi_{H7-C1-C2-H5}$). This study also shows that there is only a small difference between a treatment of the nonlocal correction at the SCF level and a perturbative correction of the results from LDA calculations.

2.2 The Ring Closures of Hexatriene and Octatetraene

The disrotatory ring closure of 1,3,5-hexatriene to 1,3-cyclohexadiene (Fig. 3) has been studied at the nonlocal BLYP/6-31G* level of theory [14]. Table 2 summarizes the results from these and from RHF/6-31G* calculations [29]. The bond length of the partial single bond C1–C6 is predicted to be 0.05 Å longer by the BLYP/6-31G* method than by the RHF/6-31G* method, whereas the other geometric parameters are found to be approximately equal by the two methods. The activation energy calculated at the BLYP/6-31G* level is in much better agreement with the experimental value of 29.0 kcal/mol than the corresponding value from the RHF/6-31G* calculation. It is also close to the value calculated at the MP2/6-31G* level [29]. The heat of reaction calculated by the RHF method is, however, closer to the experimental value [30] than the result obtained by the BLYP method.

The results for the conrotatory ring closure of octatetraene to cyclooctatriene (Fig. 4) are shown in Table 3. For this system, the results from the RHF and DFT calculations differ significantly. Using the BLYP/6-31G* method, a bond length of 2.42 Å is computed for the forming bond, which is 0.22 Å longer than the value obtained at the RHF/6-31G* level of theory. This earlier transition state obtained by the DFT method is also indicated by the smaller

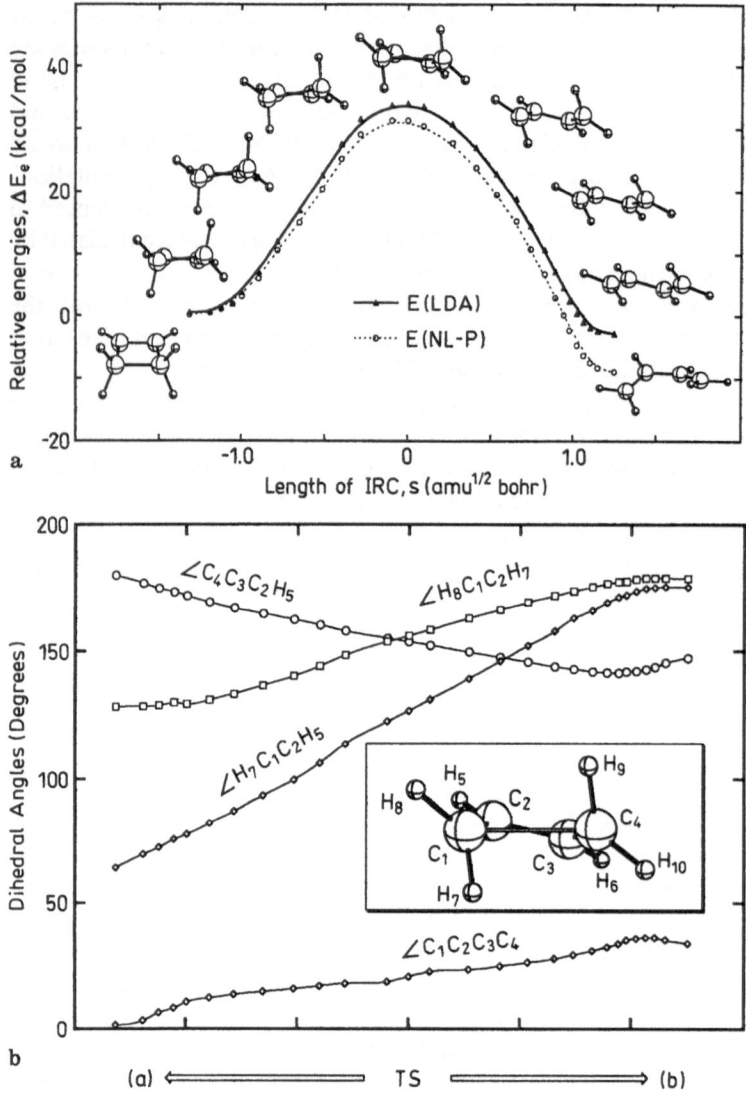

Fig. 2. IRC of the electrocyclic ring opening of cyclobutene to 1,3-butadiene. **a** Energy profile. **b** Evolution of bond distances and angles

degree of pyramidalization at the carbon termini. The activation energy calculated at the BLYP/6-31G* level agrees with the measured value of 17.0 kcal/mol [31], whereas the RHF/6-31G* method overestimates the activation energy by 16 kcal/mol [32, 33]. Inclusion of electron correlation by MP4/6-31G* calculations on MP2 geometries leads to an underestimation of activation energy by only 2 kcal/mol. The heat of reaction [34] is correctly predicted by the nonlocal DFT method, whereas the RHF/6-31G* and the MP4/6-31G* methods over-estimate the exothermicity of the reaction.

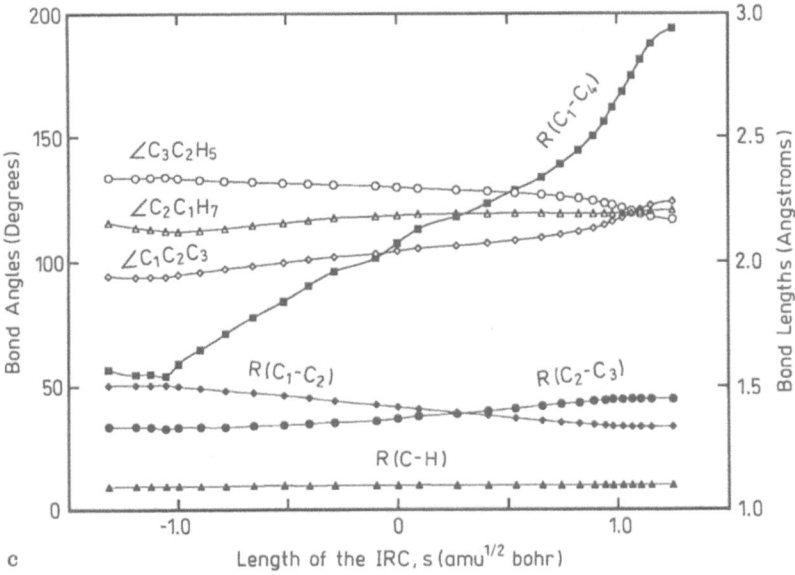

c

Fig.2c. Evolution of dihedral angles (From [12])

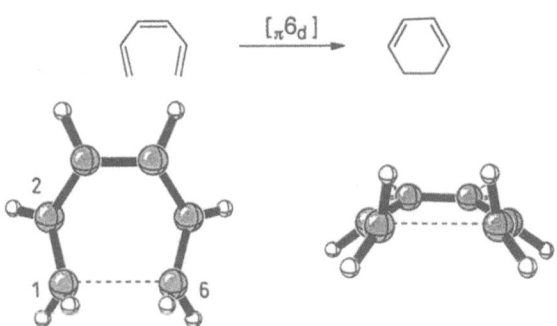

Fig. 3. Transition structure for the electrocyclic ring closure of 1,3,5-hexatriene to 1,3-cyclo-hexadiene

Table 2. Transition structure geometries, activation energies, and reaction energies for transition structures of electrocyclic ring closure of 1,3,5-hexatriene

Method	Reference	R_{C1-C6} (Å)	R_{C1-C2} (Å)	E_a (kcal/mol)	ΔE_{react} (kcal/mol)
BLYP/6-31G*	14	2.299	1.410	28.5	−8.2
RHF/6-31G*	29	2.249	1.390	45.8	−14.6
MP4SDTQ	77	—	—	30.4	—
Experimental	29	—	—	29.0	−14.5

E_a, activation energy; ΔE_{react}, reaction energy.

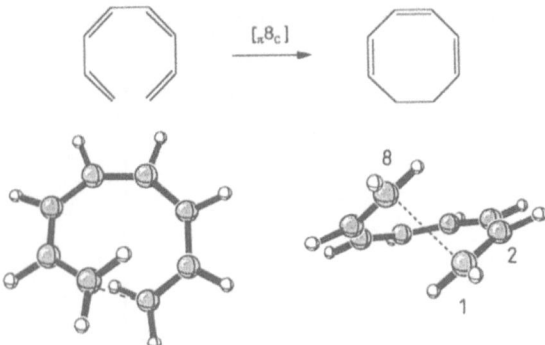

Fig. 4. Transition structure for the electrocyclic ring closure of 1,3,5,7-octatetraene to 1,3,5-cyclooctatriene

Table 3. Transition structure geometries, activation energies, and reaction energies for transition structures of electrocyclic ring closure of 1,3,5,7-octatetraene

Method	Reference	R_{C1-C8} (Å)	R_{C1-C2} (Å)	E_a (kcal/mol)	ΔE_{react} (kcal/mol)
BLYP/6-31G*	31	2.420	1.386	15.6	− 0.9
RHF/6-31G*	32	2.200	1.377	33.4	− 7.5
MP2/6-31G*	33b	2.511	1.362	12.2	− 13.5
MP4/6-31G*//MP2/6-31G*	33b	—	—	15.0	− 11.2
Experimental	31,34	—	—	17.0	− 1[a]

E_a, activation energy; ΔE_{react}, reaction energy.
[a]: For *trans,cis,cis,trans*-2,4,6,8-decatetraene [34].

3 Cycloadditions

3.1 Diels-Alder Reactions

The reaction of butadiene and ethylene has been studied at many computational levels and serves as a prototype for pericyclic reactions [6, 7]. The concerted transition structure with C_s symmetry, shown in Fig. 5, is found to be lowest in energy. The calculated activation energy varies widely, from over 45 kcal/mol at the Hartree-Fock limit to 17.6 kcal/mol, with MP2/6-31G* calculations. Inclusion of dynamic correlation energy, such as provided by QCISD(T) calculations [35], is necessary to give accurate activation energies near the experimental values of 24 kcal/mol [6, 36].

The forming C–C bond lengths found by RHF and CASSCF are predicted to be about 2.2 Å (Table 4) [5]. LDA calculations predict C–C lengths that are much too long and activation energies that are much too low. The BLYP and

Table 4. Transition structure geometries, activation energies, and reaction energies for transition structures of the Diels-Alder reaction of butadiene and ethylene

Method	Reference	R_{C1-C5} (Å)	R_{C1-C2} (Å)	R_{C2-C3} (Å)	R_{C5-C6} (Å)	$\varphi_{C3-C4-C5}$ (°C)	E_a (kcal/mol)
X_α/6-31G**	38	2.373	1.369	1.415	1.368	101.2	6.8
SVWN/6-31G**	38	2.400	1.363	1.415	1.361	100.8	3.6
BP88/6-31G**	37	2.318	1.378	1.410	1.381		18.7
BLYP/6-31G*	13	2.292	1.395	1.415	1.398	102.6	22.8
ACM/6-31G*	13	2.301	1.377	1.410	1.378	101.7	20.3
Becke3LYP/6-31G*	38	2.268	1.383	1.406	1.386	102.3	24.9
RHF/6-31G*	35	2.201	1.377	1.393	1.383	102.0	45.0
MP2/6-31G*	35	2.285	1.378	1.410	1.380	102.0	17.6
MP4SDTQ	77						22.4
CASSCF/6-31G*	35	2.223	1.398	1.397	1.404		43.8
Experimental	6,36						23.6

E_a, activation energy.

9

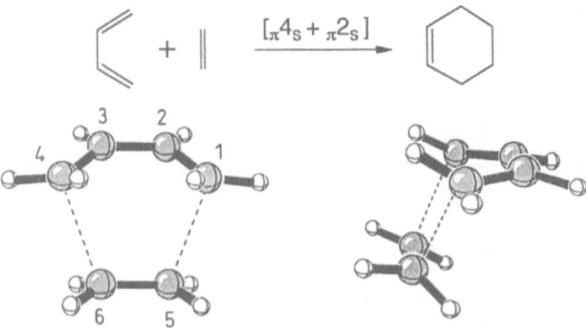

Fig. 5. Concerted transition structure of the Diels–Alder reaction of butadiene and ethylene

Becke 3LYP results are much better, both on predictions of geometry and by the more reasonable, but still too low, activation energy. Conforming to the usual pattern, the RHF activation energies are too high and MP2 are too low. Local density functional barriers are much too small, while the nonlocal and hybrid functionals according to Becke give very good activation energies [14, 37, 38].

DFT has also proved useful for the calculation of substituent effects on rates. For example, studies of reactions of butadiene and cyclopentadiene with acrolein predict a large s-*trans* preference of the acrolein, and a 1 kcal/mol lower activation energy than with the ethylene reaction, but very similar energies of *exo*- and *endo*- transition states [38]. Jursic has also studied several heterocyclic cases [39].

Of greatest significance, DFT calculations can be used to compare directly the energies of concerted and stepwise pathways. Previously, multiconfiguration SCF calculations, such as CASSCF, or configuration interaction calculations, as QCISD(T), were required to provide reasonable energies of concerted and stepwise pathways [35]. Becke 3-LYP/6-31G* calculations have been used to study the stepwise process and to compare it to the concerted mechanism. In good agreement with the results of CASSCF geometry optimizations and QCISD single points, the unrestricted Becke 3LYP/6-31G* results predict that the transition states of the stepwise mechanism involving a diradical intermediate are about 5 kcal/mol above the concerted transition state [40, 41]. Since the $< S^2 >$ values calculated for the diradicals in this and other reactions deviate considerably from the expectation value, there is some concern about the validity of these results. It has, however, been pointed out that the $< S^2 >$ values in DFT have a different physical significance than in Hartree-Fock theory [42], which is currently a topic of considerable interest [43, 44]. It is therefore not surprising that the energies obtained by DFT methods are quite reasonable at a very moderate computational cost.

3.2 1,3-Dipolar Cycloadditions

Three examples of 1,3-dipolar cycloadditions of the general type, shown in Fig. 6, have been studied using DFT methods [45]. The reaction of fulminic acid with acetylene and ethylene (Fig. 6 bottom, left and middle) and the reaction of nitrone with ethylene (Fig. 6, bottom, right) have been investigated. Table 5 summarizes the results from the DFT and selected MO calculations. In accordance with the results from RHF [46] and MCSCF calculations [47], the forming C-O bonds in the reactions of fulminic acid are predicted to be considerably longer than the C-C bonds. This asymmetry is smaller in the SVWN/DZVP calculations than in transition structures obtained by the BLYP/DZVP methods, which predict more C-C bonding. The transition states computed by the two DFT methods are much earlier than those predicted by the MO-based methods. This relationship is reversed in the case of the reaction of nitrone and ethylene. Whereas in the transition structures calculated by the MO-based methods and the SVWN method the C-C bond is longer than the C-O bond, the BLYP/method predicts the C-C bond to be shorter.

The local density approximation yields only poor estimates of the activation energies for the three model systems studied. The transition structures are calculated to be more stable than the corresponding reactants. This error is corrected by the nonlocal corrections. Although the activation energies obtained with SVWN or BLYP functionals are still much lower than the results from highly correlated MCSCF calculations, they are close to the results from MP4SDTQ/6-311G**//RHF/6-311G** calculations [48] and experimental estimates [49]. The SWVN method greatly overestimates the heat of reaction for the 1,3-dipolar cycloadditions; the inclusion of nonlocal corrections lowers the calculated heats of reaction by 24–26 kcal/mol.

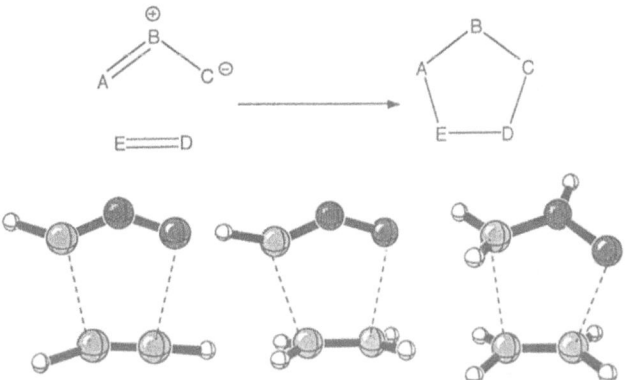

Fig. 6. Transition structures for the 1,3-dipolar cycloadditions of fulminic acid with acetylene (*left*) and ethene (*middle*) and of nitrone with ethylene (*right*)

Table 5. Transition structure geometries, activation energies, and reaction energies for transition structures of 1,3-dipolar cycloadditions

Method	Reference	Fulminic acid + acetylene				Fulminic acid + ethylene				Nitrone + ethylene			
		R_{C-C} (Å)	R_{C-O} (Å)	E_a (kcal/mol)	ΔH (kcal/mol)	R_{C-C} (Å)	R_{C-O} (Å)	E_a (kcal/mol)	ΔH (kcal/mol)	R_{C-C} (Å)	R_{C-O} (Å)	E_a (kcal/mol)	ΔH (kcal/mol)
SVWN/DZVP	45	2.436	2.505	−1.5	−108.7	2.423	2.543	−2.4	−63.4	2.335	2.242	−3.5	−49.1
BLYP/DZVP	45	2.295	2.573	9.3	−82.8	2.287	2.521	8.5	−39.6	2.177	2.230	12.9	−23.3
RHF/6–311G**	46	2.139	2.185	36.8	−82.0	2.142	2.228	34.5	−49.3	2.191	2.040	29.5	−40.5
MP2/6–311G**[a]	46	—	—	12.0	−79.2	—	—	9.6	−43.2	—	—	7.0	−39.6
MCSCF/4–31G	47	2.17	2.28	26.0	−66.9	2.08	2.32	28.8	−27.4	2.15	2.23	24.6	−21.2

E_a, activation energy.
[a], On RHF/6–311G** geometry.

Recently, Sustmann has applied DFT to the 1,3-dipolar cycloadditions of the parent nitrone to ethylene and thioformaldehyde. The results were also compared to MP2/6-31G* optimization and QCISD(T) energetics with good success [50].

4 Sigmatropic Shift Reactions

4.1 The Vinylcyclopropane-Cyclopentene Rearrangement

The rearrangement of vinylcyclopropane to form cyclopentene (Fig. 7) is the most simple [1, 3] carbon shift observed experimentally. This reaction has excited considerable interest during the last decade not only because of its intriguing mechanism and the relationship of stereochemistry to the Wood-ward-Hoffmann rules, but also because of the synthetic utility of a variety of substituted cases [51]. The diradicaloid stepwise path (upper part of Fig. 7) seems to be at least competitive with the concerted, symmetry allowed [1a, 3s] reaction (lower part of reaction scheme). The experimental results are, however, not unequivocal. Whereas several secondary kinetic isotope effects (SKIE) studies [52–54] can be interpreted by a stepwise mechanism, stereochemical studies show a high degree of stereoselectivity for the suprafacial-inversion pathway predicted for a concerted pathway [55–57]. In addition, vinylcyclop-ropanes undergo rapid stereomutation which presumably also involves a di-radicaloid pathway [58], complicating stereochemical studies.

The vinylcyclopropane rearrangement has been studied using a variety of semiempirical methods [59, 60], including a study of the reaction dynamics at the AM1 level [61]. Recently, both possible pathways have been studied using various MO and DFT methods [62]. The results for the two possible pathways are shown in Table 6.

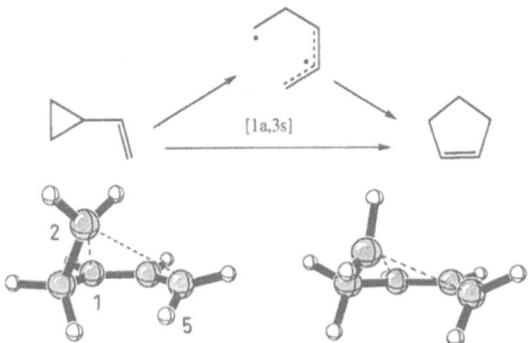

Fig. 7. The [1,3] sigmatropic shift of vinylcyclopropane and two transition structures involved in the reaction. *Left* transition structure for the concerted process; *right* transition structure for the bond closure in the diradical pathway

Table 6. Transition structure geometries, activation energies, and reaction energies for transition structures of vinylcyclopropane to cyclopentene rearrangement [62]

Method	Concerted				Stepwise			
	R_{C1-C2} (Å)	R_{C2-C5} (Å)	E_a (kcal/mol)	ΔH (kcal/mol)	R_{C1-C2} (Å)	R_{C2-C5} (Å)	E_a (kcal/mol)	ΔH (kcal/mol)
Becke3LYP/6–31G*	2.284	2.637	59.0	– 20.7	2.489	2.680	49.0	– 20.7
HF/6–31G*	2.214	2.555	82.6	– 21.8	2.430	2.409	28.5	– 21.8
MP2/6–31G*	2.281	2.610	59.8	—	2.415	2.576	34.7	—
MCSCF (4e/4o)6–31G*	—	—	—	—	2.451	2.810	17.9	—

E_a, activation energy.

As already noted earlier, the transition structure geometry and activation energies for the concerted reaction pathway obtained by the Becke 3LYP functional are very similar to the results from the MP2/6-31G* calculations. The bond lengths predicted by the RHF/6-31G* method, on the other hand, are ~ 0.6 Å shorter than either the Becke 3LYP or the MP2 results. Although the activation energies calculated by these two methods are ~ 23 kcal/mol lower than the one obtained from the RHF/6-31G* calculations, they are still too high by about 10 kcal/mol as compared with the experimental value of 49.7 kcal/mol [63], indicating an energetic preference for the diradicaloid mechanism.

At the UHF/6-31G* level of theory, the ring closure of the diradicaloid intermediate to form cyclopentene has been found to be the rate determining step. This confirms an earlier analysis of the experimental SKIE by Gajewski et al. [53]. The results of the calculations for the transition structure for ring closure in the stepwise pathway using different methods are also summarized in Table 6. For the breaking bond C1–C2, a comparably small range of bonding distances ranging from 2.42 Å to 2.49 Å is calculated by the four methods used. For the forming bond C2–C5, the range of the computed bond lengths is considerably larger, between 2.41 Å, calculated by the HF/6-31G* method, and 2.81 Å, obtained by MCSCF (4e/4o)/6-31G* calculations. The Becke 3LYP/6-31G* methods also predict a very early transition state for bond closure, with a bond distance of 2.68 Å.

The activation energy computed by the Becke 3LYP/6-31G* method is in excellent agreement with the experimental value, whereas the activation energy values computed by other methods are too low. This is presumably because of the use of a UHF wavefunction and the overestimation of the stability of the diradical by the MP2 and MCSCF methods. It has been found in a number of earlier studies [64–66] that the $< S^2 >$ values calculated by DFT methods are usually very close to the expected value. This is also found in the case of the bond closure transition state, where the $< S^2 >$ value computed by the Becke 3LYP/6-31G* and the UHF/6-31G* methods are 0.17 and 1.05, respectively. This example provides another example which shows that appropriate DFT methods are useful for the study of competition between concerted and stepwise mechanisms.

Since the theoretical SKIE calculated for the ring closure transition structure reproduces the experimental SKIE accurately, it can be deduced that the vinylcyclopropane-cyclopentene rearrangement proceeds via a diradicaloid stepwise mechanism.

4.2 The [1,5]-*H* Shift in Pentadiene

The degenerate [1,5] hydrogen shift reaction of (Z)-1,3-pentadiene has been studied by a number of ab intio [67–69] and semiempirical [70–72] methods. Since the energy of concert [73] for this reaction is approximately 40 kcal/mol, there is no doubt that this reaction has a concerted mechanism with an aromatic transition state [74] (Fig. 8). The bond lengths in the transition structures calculated by the different methods deviate from each other only by several hundredths of an Å [6]. These findings are similar to the results for the electrocyclic ring opening of cyclobutene (Sect. 2.1) in that the geometries obtained by various methods are very similar, but higher levels of theory are required for an accurate prediction of the activation energy.

Table 7 shows the results from DFT and selected MO calculations. The geometries obtained by the local and nonlocal DFT methods are very similar to the ones calculated by MO-based methods. The LDA calculations, however, underestimate the activation energy by 10–13 kcal/mol. The use of nonlocal gradient corrections improves the performance of the calculations, although the activation energy is still underestimated by 3–5 kcal/mol. Unfortunately, no calculations using hybrid methods have been published yet.

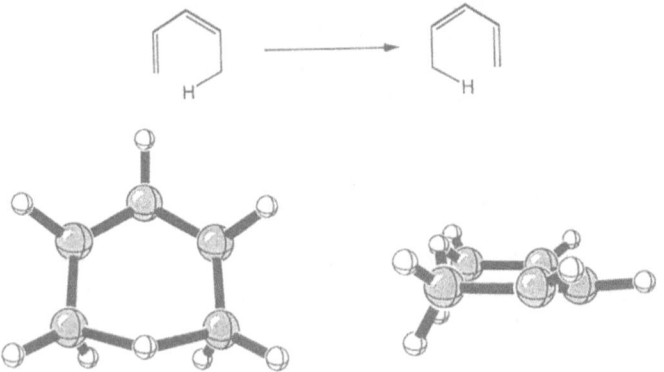

Fig. 8. Transition structure of the [1, 5] hydrogen shift in (Z)-1,3- pentadiene

Table 7. [1,5] Transition structure geometries, activation energies, and reaction energies for transition structures of [1,5]-hydrogen shift in pentadiene

Method	Reference	R_{C1-C2} (Å)	R_{C1-H} (Å)	E_a (kcal/mol)
LDA/DZVP	16	1.418	1.410	23.6
LDA/6–31G**	37	1.409	1.414	26.0
BP/DZVP	16	1.415	1.426	30.9
BP/6–31G**	37	1.427	1.425	32.9
RHF/6–31G**	67	1.405	1.440	59.8
MP2/6–31G*	74	1.418	1.411	36.5
RMP4SDTQ/6–311G**//	74			34.9
QCISD(T)	77			36.9

E_a, activation energy.

4.3 The Cope Rearrangement

The precise mechanism of the degenerate [3, 3]-sigmatropic shift reaction of 1,5-hexadiene has been the subject of a fierce debate over the last two decades [7] (Fig. 9). The question of whether the reaction has an aromatic-type or a 1,4-diyl-type transition state or even a short-lived 1,4-diyl intermediate has been studied with a variety of theoretical methods. The third possibility, a bisallylic pathway, has been ruled out based on energetics [75]. Semiempirical [76] and MP2 [77] calculations predict a tight, 1,4-diyl structure. A harmonic frequency analysis of these structures shows that they correspond to minima on the potential energy surface. Recent CASSCF calculations with inclusion of second-order perturbation theory corrections yield a tight transition structure with very little diradical character [78–79], but RHF [80] and CASSCF [81] calculations yielded loose, aromatic-type transition structures. The possibility of several interconnected transition structures has also been proposed [82], demonstrating that the potential energy surface in the transition state region is extremely flat. Since the differences in activation energies calculated for transition structures with a wide range of bond distances for the forming and breaking bonds are very small, additional criteria are needed for the validation of the theoretical results.

Table 8 summarizes the results of the calculations of the C_{2h} symmetric chair and the C_{2v} symmetric boat transition structures [83]. The local spin density approximation predicts a tight transition structure, comparable to the one obtained by the MP2(fc)/6-31G* method, but fails to reproduce the experimental activation energies [84]. The use of the gradient-corrected BLYP functional yields loose, aromatic-type transition structures and improved activation energies for the chair transition structures. The activation energy for the boat transition structure is, however, too low by 9–10 kcal/mol as compared to the experimental value of 44.7 kcal/mol [85]. The activation energies for the chair and the boat transition structures obtained by the Becke 3LYP method are in

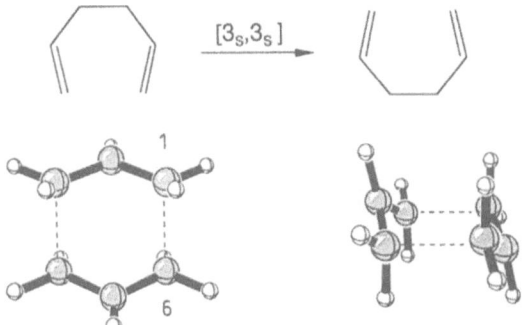

Fig. 9. Transition structure of the Cope rearrangement of 1,5 hexadiene

Table 8. Transition structure geometries, activation energies, and reaction energies for transition structures of Cope rearrangement of 1,5-hexadiene

Method	Reference	Chair		Boat	
		R_{C1-C6} (Å)	E_a (kcal/mol)	R_{C1-C6} (Å)	E_a (kcal/mol)
SVWN/6–31G**	83	1.753	19.8	1.966	33.2
BLYP/6–31G*	83	2.034	29.7	2.289	36.0
BLYP/6–311 + G**	83	2.145	30.0	2.376	35.0
Becke3LYP/6–31G*	83	1.971	34.2	2.208	42.0
Becke3LYP/6–311 + G*	83	2.007	32.2	2.248	39.6
Becke3LYP/6–311 + G**	83	2.043	34.8	2.279	41.2
RHF/6–31G*	80	2.046	56.6	2.203	
MP4SDTQ/6–31G*//		1.794	32.1	2.055	43.1
MP2(fc)/6–31G*	77				
CASSCF/6–31G*	82	2.189	47.7	2.615	
CASPT2N/6–311G(2d,2p)	78	1.885	32.2	2.204	41.2
QCISD(T)	77		35.8		45.0
Experimental	84,85	—	33.3 ± 0.5	—	44.7 ± 2.0

E_a, activation energy.

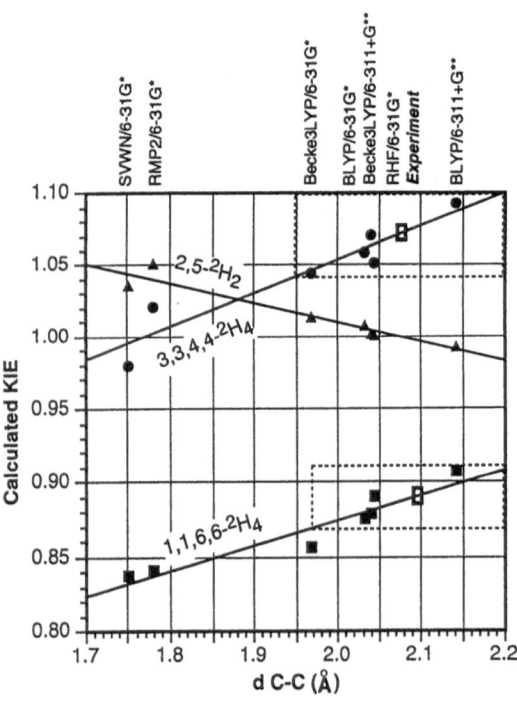

Fig. 10. Calculated secondary kinetic isotope effects versus interallylic distance, R_{C1-C6}. *Rectangular dashed boxes* represent the experimental error limits. (From [82])

much better agreement with the experimental value. The interallylic distance in the transition structure is smaller than the one computed by the BLYP method but is still characteristic of an aromatic-type transition state.

As already pointed out, the energies obtained by the different basis sets used are quite similar, despite the large differences in the calculated interallylic distances. Since SKIE are a very sensitive probe for transition state geometries, the relationship between calculated bond lengths for the forming/breaking bond and the theoretical SKIE has been analyzed [82]. Fig. 10 shows a plot of the interallylic distance versus the theoretical SKIE. Independent of the methods and basis sets used, the plot shows the good correlation between the two parameters. Based on the experimental SKIE [86, 87], the interallylic distance in the transition state can be estimated to be 2.085 ± 0.125Å.

4.4 The Claisen Rearrangement

A reaction closely related to the Cope rearrangement is the [3, 3] sigmatropic rearrangement of allyl vinyl ether to form 4-pentenal (Fig. 11). The results of DFT and selected MO-based calculations are summarized in Table 9. As for the Cope rearrangement, a tight transition state is predicted by the SVWN method, which also underestimates the activation energy for the Claisen rearrangement. Although the BLYP/6-31G* method yields a much looser transition structure, the calculated activation energy is similar to the one computed by the SVWN method. The lengths of the forming ($R_{C_1-C_6}$) and breaking ($R_{O_3-C_4}$) bonds in the transition structure obtained by the Becke 3LYP functional are intermediate between the RHF/6-31G* [88] and the CASSCF(6e/6o)/6-31G* [89] results. The Becke 3LYP calculations give a much better activation energy of 26.8 kcal/mol, close to the experimental activation energy of 30.8 kcal/mol [90, 91].

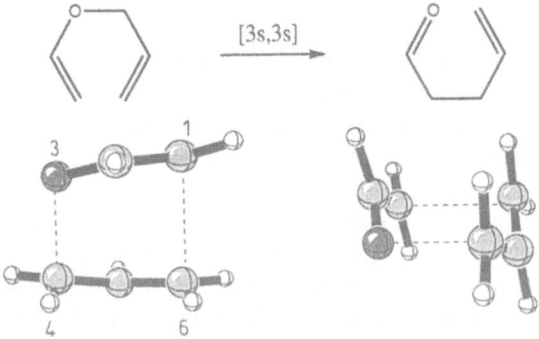

Fig. 11. Transition structure for the Claisen rearrangement of allyl vinyl ether to 4-pentenal

Table 9. Transition structure geometries, activation energies, and reaction energies for transition structures of Claisen rearrangement of allyl vinyl ether

Method	Reference	R_{C1-C6} (Å)	R_{O3-C4} (Å)	E_a (kcal/mol)
SVWN/6–31G*	83	2.094	1.696	20.8
BLYP/6–31G*	83	2.428	1.959	21.1
Becke3LYP/6–31G*	83	2.312	1.902	26.8
Becke3LYP/6–311 + G**	83	2.384	1.954	26.1
RHF/6–31G*	88	2.266	1.918	47.7
CASSCF/6–31G*	89	2.564	2.100	42.5
Experimental	90,91			30.8

E_a, activation energy.

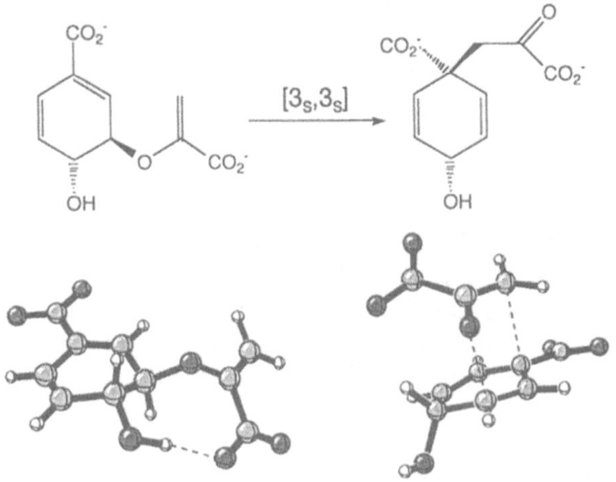

Fig. 12. Pseudodiequatorial reactant (*left*) and transition structure (*right*) for the Claisen rearrangement of chorismate

    The Claisen rearrangement has attracted special attention because of the pronounced solvent dependence of the reaction [92] and the biochemically important Claisen rearrangement of chorismate to prephenate in the shikimic acid pathway [93] (Fig. 12). Both aspects of the reaction have been studied recently using DFT methods.

    The effects of solvation have been studied by an implementation of an ellipsoid cavity model into Kohn-Sham theory [94]. The lowering of the activation energy by water solvation of the allyl vinyl ether has been calculated to be 0.3 kcal/mol by this method. This value is considerably lower than the results from other calculations on the same system using cavity models [95], free energy perturbation [96], or QM/MM calculations [97] as well as the values deduced from the rate acceleration observed for a number of substituted allyl vinyl ethers [98–99]. The Claisen rearrangement of chorismate, chorismic acid, and numerous model systems has been studied by the BLYP/6-31G* method [100]. The transition structures obtained by these calculations are much looser than the ones obtained by RHF calculations [101–103].

5 Conclusions

This review has provided an overview of the studies of pericyclic reaction transition states using density functional theory methods up to the middle of 1995. Since the parent systems for most of the pericyclic reaction classes have been studied, a first assessment of DFT methods for the calculation of pericyclic transition structures can be made.

It is clear that the local density approximations (such as X_α or SVWN) give reasonable estimates of the geometries of the reactants. However, transition state geometries and activation energies are generally not in good agreement with the data from other sources. The activation energies are systematically highly underestimated, and the relative energies obtained for reactants and products are unreliable. Therefore, these methods are not appropriate for the study of organic reaction mechanisms.

The methods using nonlocal gradient corrections (e.g., BLYP, BP) offer significant improvements. The bond lengths calculated are in general slightly longer than the ones obtained by other high-level calculations whereas the activation energies seem to be slightly underestimated as compared to the available experimental data. This is in agreement with the findings of performance studies for other reaction types [104–106].

It appears that the hybrid methods based on Becke's three parameter exchange correlation functional (Becke3LYP, ACM) yield the most reliable results for geometries and energies [107, 108]. The values obtained in these calculations are very close to high-level MO-based calculations and are in good agreement with the available experimental data at a computational cost comparable to Hartree-Fock calculations. It has also been pointed out [109] that basis set convergence in DFT is much faster than in MO-based methods, therefore requiring smaller basis sets for high-level calculations. This is confirmed by the studies of pericyclic reactions, where there is generally only a small difference between double zeta and triple zeta basis sets.

Although mostly parent cases have been studied so far, it can be expected that because of these improvements in speed and accuracy it is likely that these methods will be increasingly used for the study of chemically important substituted cases. The current intense investigations of new functionals as well as the exploration of the limitations of density functional theory will yield an even better understanding of the advantages and disadvantages of density functional theory. It can, however, be stated that, even at this point, DFT methods provide a useful alternative to conventional Hartree-Fock methods for the study of pericyclic reactions.

References

1. Woodward RB, Hoffmann R (1969) Angew Chem 81:797; Angew Chem Int Ed Engl 8:781
2. Fukui K (1965) Tetrahedron Lett 2009
3. Fukui K (1965) Tetrahedron Lett 2427
4. Houk KN (1975) Acc Chem Research 8:361
5. Fleming I (1977) Frontier orbitals and organic chemical reactions Wiley, New York
6. For generalizations and a review, see: Houk KN, Li Y, Evanseck JD (1992) Angew Chem 104: 711; Angew Chem Int Ed Engl 31:682
7. Houk KN, Gonzalez J, Li Y (1995) Acc Chem Research 28:81
8. Eg: Breulet F, Schaefer III HF (1984) J Am Chem Soc 106:1221
9. Spellmeyer DC, Houk KN (1988) J Am Chem Soc 110:3412
10. Dewar MJS, Zoebisch EG, Healy EF, Stewart JJP (1985) J Am Chem Soc 107:3902
11. Hsu K, Buenker RJ, Peyerimhoff SD (1971) J Am Chem Soc 93:2117
12. Deng L, Ziegler T (1995) J Phys Chem 99:612
13. Baker J, Muir M, Andzelm J (1995) J Chem Phys 102:2063
14. Wiest O, Houk KN, Black KA, Thomas IV B (1995) J Am Chem Soc 117:8594
15. Wiest O, Houk KN unpublished results
16. Carpenter JE, Sosa CP (1994) J Mol Struct (Theochem) 311:325
17. Cooper W, Walters WD (1958) J Am Chem Soc 80:4220
18. Carr RW, Walters WD (1965) J Phys Chem 69:1073
19. Gajewski JJ (1981) Hydrocarbon thermal isomerizations. Academic Press, New York, p. 49
20. Lipnick RL, Garbisch EW (1973) J Am Chem Soc 95:6370
21. Furukawa Y, Takeuchi H, Tasumi M (1983) Bull Chem Soc Jpn 56:392
22. Vosko SH, Wilk L, Nussair M (1980) Can J Chem 58:1200
23. The basis sets used in these calculations have been optimized for SVWN calculations: Godbout N, Salahub DR, Andzelm J, Wimmer E (1992) Can J Chem 72:560
24. Becke AD (1993) J Chem Phys 98:5648
25. Becke AD (1988) Phys Rev A38:3098
26. Lee C, Yang W, Parr RG (1988) Phys Rev B 37:785
27. Perdew J (1986) P Phys Rev B33:8822; compare also: Perdew JP (1986) Phys Rev B34:7406
28. Deng L, Ziegler T (1994) Int J Quant Chem 52:731
29. Baldwin JE, Reddy VP, Hess Jr BA, Schaad LJ (1988) J Am Chem Soc 110:8554
30. Lewis KE, Steiner H (1964) J Chem Soc 3080
31. Goldfarb TD, Landquist LJ (1967) J Am Chem Soc 89:4588
32. Thomas IV BE, Evanseck JD, Houk KN (1993) Isr J Chem 33:287
33a. Thomas IV BE, Evanseck JD, Houk KN (1993) J Am Chem Soc 115:4165
33b. Jiao H, Schleyer PvR (1994) J Chem Soc Perkin Trans 2:407
34. $\Delta H_{React} = -1$ kcal/mol for *trans, cis, cis, trans*-2,4,6,8-decatetraene: Huisgen R, Dahmen A, Huber H (1969) Tetrahedron 1461
35. Li Y, Houk KN (1993) J Am Chem Soc 115:7478
36. Uchiyama M, Tomioka T, Amano A (1964) J Phys Chem 68:1878
37. Stanton RV, Merz Jr KM (1994) J Chem Phys 100:434
38. Jursic B, Zdravkovski Z (1995) J Chem Soc Perkin Trans 2 1223
39. Jursic B (1995) J Org Chem 60:4721
40. Goldstein E, Beno B, Houk KN (1995) J Am Chem Soc, in press

41. Ochoa C, Beno B, Houk KN (1995) submitted for publication
42. Erikson IA, Malikina OL, Malkin VG, Salahub DR (1994) J Chem Phys 1994 100:5066
43. Wang J, Becke AD, Smith Jr VH (1995) J Chem Phys 102:3477
44. Laming GJ, Handy NC, Amos RD (1993) Mol Phys 80:1121
45. Sosa C, Andzelm J, Lee C, Blake JF, Chenard BL, Butler TW (1994) Int J Quant Chem 49:511
46. Brown FK, Singh UC, Kollmann PA, Raimondi L, Houk KN, Bock CW (1992) J Org Chem 57:4862
47. McDoall JJW, Robb MA, Niazi U, Bernardi F, Schlegel B (1987) J Am Chem Soc 109:4642
48. Fulminic acid + acetylene: $E_a = -11.0$ kcal/mol, $\Delta H = -77.7$ kcal/mol; Fulminic acid + ethylene: $E_a = 9.5$ kcal/mol; $\Delta H = -42.8$ kcal/mol
49. Firestone RA (1977) Tetrahedron 33:3009
50. Sustmann R, Sicking W, Huisgen R (1995) J Am Chem Soc 117:9679
51. Hudlicky T, Kutchan TM, Naqvi SM (1985) Organic Reactions 33:247
52. Chickos J (1984) National ACS Meeting St Louis ORG228
53. Gajewski JJ, Olson LP (1991) J Am Chem Soc 113:7432; Gajewski JJ, Olson LP, Willcott MR (1996) J Am Chem Soc 118:299
54. Baldwin JE, Villarica KA (1994) Tetrahedron Lett 35:7905
55. Gajewski JJ, Squicciarini MP (1989) J Am Chem Soc 111:6717
56. Baldwin JE, Bonacorsi S (1994) J Org Chem 59:7401
57. Baldwin JE, Villarica KA, Freedberg DI, Anet FAL (1994) J Am Chem Soc 116:10845
58. Wilcott MR III, Cargle VH (1969) J Am Chem Soc 91:4310
59. Dewar MJS, Fonken GJ, Kirschner S, Mitner DE (1975) J Am Chem Soc 97:6750
60. Quirante JJ, Enriquez F, Hernando JM (1990) J Mol Struct (THEOCHEM) 204:193
61. Carpenter BK (1995) J Am Chem Soc 117:6336
62. Wiest O, Houk KN, Storer J, Fennen J, Nendel M (1995) National ACS Meeting Chicago ORGN33
63. Wellington CA (1962) J Phys Chem 66:1671
64. Adamo C, Barone V, Fortunelli A (1994) J Phys Chem 98:8648
65. Yamanaha S, Kawakami T, Nagao H, Yamaguchi K (1994) Chem Phys Lett 231:25
66. Baker J, Scheiner A, Andzelm J (1993) Chem Phys Lett 216:380
67. Jensen F, Houk KN (1987) J Am Chem Soc 109:3139
68. Kahn DD, Hehre WJ, Rondan NG, Houk KN (1985) J Am Chem Soc 107:8192
69. Rondan NG, Houk KN (1984) Tetrahedron Lett 25:2519
70. Dewar MJS, Healy EF, Ruiz JM (1988) J Am Chem Soc 110:2666
71. Kahn DD, Hehre WJ, Rondan NG, Houk KN (1985) J Am Chem Soc 107:8192
72. Liu Y-P, Lynch GC, Truong TN, Lu D-H, Truhlar DG, Garrett BC (1993) J Am Chem Soc 113:2408
73. Doering WvE, Roth WR, Breuckmann R, Figge L, Lennartz H-W, Fessner WD, Prinzbach H (1988) Chem Ber 121:1
74. The aromaticity of the calculated transition structure has also been confirmed by calculation of the magnetic susceptibility: Jiao H, Schleyer PvR (1994) J Chem Soc Faraday Trans 90:1559
75. Roth WR, Bauer F, Beitat A, Ebbrecht T, Wüstfeld M (1991) Chem Ber 124:1453
76. Dewar MJS, Jie C (1992) Acc Chem Res 25:537
77. Jiao H, Schleyer PvR (1995) Angew Chem 107:329 Angew Chem Int Ed Engl 34:334
78. Hrovat DA, Morokuma K, Borden WT (1994) J Am Chem Soc 116:1072
79. Kozlowski PM, Dupuis M, Davidson ER (1995) J Am Chem Soc 117:774
80. Borden WT, Loncharich RJ, Houk KN (1988) Ann Rev Phys Chem 39:213
81. Houk KN, Gustafson SM, Black KA (1992) J Am Chem Soc 114:8565
82. Dupuis M, Murray C, Davidson ER (1991) J Am Chem Soc 113:9756
83. Wiest O, Black KA, Houk KN (1994) J Am Chem Soc 116:10336
84. Doering WvE, Toscano VG, Beasley GH (1971) Tetrahedron 27:5299
85. Doering WvE, Roth W (1962) Tetrahedron 18:67
86. For early studies on SKIE of the Cope reaction see: Humski K, Malojcic R, Borcic S, Sunko DE (1970) J Am Chem Soc 92:6534

87. Experimental kinetic isotope effects are for the 1-methyl and 3-methyl 1,5-hexadienes: Gajewski JJ, Conrad ND (1979) J Am Chem Soc 101:6693
88. Vance RL, Rondan NG, Houk KN, Jensen F, Borden WT, Kormonicki A, Wimmer E (1988) J Am Chem Soc 110:2314
89. Yoo H-Y, Houk KN (1994) J Am Chem Soc 116:12047
90. Schuler FW, Murphy GW (1950) J Am Chem Soc 72:3155
91. The value of 30.8 kcal/mol has been criticized in the literature: Coates RM, Rodgers BD, Hobbs SJ, Peck DR, Curran DP (1987) J Am Chem Soc 109:1160
92. White WN, Wolfarth EF (1970) J Org Chem 35:2196
93. Haslam E (1993) Shikimic Acid Metabolism and Metabolites Wiley, New York
94. Davidson MM, Hillier IH, Hall RJ, Burton NA (1994) J. Am Chem Soc 116:9294
95. Storer JW, Giesen DJ, Hawkins GD, Lynch GC, Cramer CJ, Truhlar DG (1994) in: Cramer CJ, Truhlar DG (ed) Structure and Reactivity in Aqueous Solutions pp.24–49 (ACS Symposium Series 568). Other contributions to the same volume discuss alternative computational approaches to the solvent dependency of the Claisen reaction
96. Severance DL, Jorgensen WL (1992) J Am Chem Soc 114:10966.
97. Gao J (1994) J Am Chem Soc 116:1563
98. Gajewski JJ, Jurazj J, Kimbrough DR, Grande ME, Ganem B, Carpenter BK (1981) J Am Chem Soc 103:6983
99. Brandes E, Grieco PA, Gajewski JJ (1989) J Org Chem 54:515
100. Wiest O, Houk KN, J Am Chem Soc (1995) 117:11628
101. Davidson MM, Hillier IH (1994) J Chem Soc Perk Trans 2 1415
102. Davidson MM, Hillier IH (1994) Chem Phys Lett 225:293
103. Wiest O, Houk KN (1994) J Org Chem 59:7582
104. Andzelm J, Wimmer E (1992) J Chem Phys 96:1280
105. Johnson BG, Gill PMW, Pople JA (1993) J Chem Phys 98:5612
106. Ziegler T (1991) Chem Rev 91:651
107. Oliphant N, Bartlett RJ (1994) J Chem Phys 100:6550
108. Gill PMW, Johnson BG, Pople JA, Frisch MJ (1992) Int J Quant Chem: Quant Chem Symp 26:319
109. Rauhut G, Pulay P (1995) J Phys Chem 99:3093

Reactivity Criteria in Charge Sensitivity Analysis*

Roman F. Nalewajski[1], Jacek Korchowiec[1], and Artur Michalak[2]

[1]K. Gumiński Department of Theoretical Chemistry, and [2]Department of Computational Methods in Chemistry, Jagiellonian University, R. Ingardena 3, 30-060 Cracow, Poland

Table of Contents

*This work has been supported by the Commission of European Communities.

The charge sensitivity analysis, based upon the hardness/softness concepts and the chemical potential (electronegativity) equalization principle established within the density functional theory, is used as a diagnostic tool for probing trends in the chemical reactivity of large molecular systems. The new criteria are reviewed with special emphasis on two-reactant reactivity concepts, which explicitly take into account the interaction between reactants in a general donor-acceptor system, and the collective charge displacement coordinate systems that give the most compact description of the charge reorganization accompanying chemical reactions. The global collective populational reference frames discussed include populational normal modes, minimum energy coordinates, and the relaxational modes; the reactant reference frames include internal modes of reactants as well as externally decoupled and inter-reactant-coupling modes. The charge-coupling information in the atoms-in-molecules resolution is modelled by the hardness matrix, which provides the canonical input data for determining a series of chemically interesting probes for diagnosing reactivity trends. A survey of these concepts includes the global treatment of molecular systems and the reactant-resolved description of general reactive systems. The molecular-fragment development emphasizes the relaxational influence of one reactant upon another, reflected by the off-diagonal charge sensitivities that measure the responses of one reactant to charge displacements in the other. Both open (exchanging electrons with the reservoir) and closed (preserving the number of electrons) reactive systems are investigated. Stability criteria for the equilibrium charge distribution in such systems are summarized and their implications discussed. The ground-state mapping relations between geometrical (nuclear position) and electronic (atomic charge) degrees of freedom are related to the Gutmann rules of structural chemistry. Illustrative examples are given of the application of these reactivity concepts to model catalytic clusters of transition metal oxide surfaces and large adsorbates (toluene).

List of Acronyms

A	Acid, Acidic (Reactant)
AIM	Atoms-in-Molecules (Level of Resolution)
B	Base, Basic (Reactant)
CI	Configuration Interaction (Method)
CS	Charge Sensitivities
CSA	Charge Sensitivity Analysis
CT	Charge Transfer (External or Internal)
DFT	Density Functional Theory
E	External (Stability/Instability)
EDM	Externally Decoupled Modes
EDM ($\rightarrow$ IDM)	EDM Determined to Resemble the Most (MOC) IDM
EDM ($\rightarrow$ MEC)	EDM Determined to Resemble the Most (MOC) MEC
EEM	Electronegativity Equalization Method
EPS	Electron Population Space
FF	Fukui Function
G	Group (Level of Resolution)
g	Global (Level of Resolution)
HK	Hohenberg-Kohn (DFT)
HOMO	Highest Occupied Molecular Orbital (Frontier MO)
I	Internal (Stability/Instability)
IDM	Internally Decoupled Modes
INDO	Intermediate Neglect of Differential Overlap (Approximation, Method)
INM	Internal Normal Modes
IRM	Inter-Reactant Modes
L	Local (Level of Resolution)
LUMO	Lowest Unoccupied Molecular Orbital (Frontier MO)
M	Molecular System
m-	Meta- Position in the Substituted Benzene Ring
MEC	Minimum Energy Coordinates
MNDO	Modified Neglect of Differential Overlap [NDDO (Neglect of Diatomic Differential Overlap) Approximation, Method]
MO	Molecular Orbital (Theory, Level of Resolution)
MOC	Maximum Overlap Criterion
NPS	Nuclear Position Space
o-	Ortho- Position in the Substituted Benzene Ring
P	Polarization/Polarizational (Internal or External)
p-	Para- Position in the Substituted Benzene Ring
PNM	Populational Normal Modes

REC   Relaxational Coordinates
REM   Relaxational Modes
SCF   Self Consistent Field (Method)
SINDO1   Symmetrically Orthogonalized INDO (Method)
STO-3G   Minimum Basis Set of Slater-type Orbitals, each Represented by 3 Gaussian Orbitals

1 Introduction

The *density functional theory* (DFT) [1–5] has provided a framework for formulating new approaches to classical problems in chemistry: the origin of chemical bonding, factors determining the selectivity and relative importance of alternative reaction sites and pathways in large systems, stability of molecular charge distribution, treatments of open and closed molecular systems, etc. [3, 6–8]. Some of the basic DFT concepts were shown to be related to fundamental chemical concepts, *viz.*, electronegativity and hardness of the molecular charge distribution [3]. Recently, a variety of charge sensititivies, both local and nonlocal, and the related geometrical reactivity concepts, have been defined at different levels of resolution, e.g., the hardnesses and softnesses of functional groups and reactants, hardness and softness matrices and kernels, the Fukui function (FF) indices, response quantities involving external potentials, etc. [3, 6–10]. A systematic procedure called *Charge Sensitivity Analysis* (CSA) [7, 9] has been developed for determining chemically interesting charge responses that can be used to probe and diagnose trends in chemical reactivity at different stages of a reaction. Since the *Electronegativity Equalization Method* (EEM) [10] represents a particular realization of CSA, we shall jointly refer to those approaches as CSA/EEM.

One of the major goals of theoretical chemistry is to identify the various electronic and geometrical factors that influence chemical reactions, in order to probe chemical reactivity trends of large molecular systems, for which exact *ab initio* calculations are prohibitively expensive, and to formulate the desired, favourable "*matching relations*" between reactants in terms of the appropriate *reactivity criteria*. The so called charge sensitivities represent the second partials of the system energy with respect to the electron population (density) and/or the external (chemical) potential variables, as well as the corresponding complementary inverse quantities. They provide a sound, novel basis for formulating *reactivity indices*, rooted in the corresponding expressions for the interaction energy, which can be used to both explain and predict trends in chemical reactivity, as well as to formulate and rationalize general *reactivity rules* for various types of chemical processes. Since charge sensitivities carry the information about the response of the system to displacements in both the external potential (v) and the global number of electrons (N) or the electron populations of constituent atoms ($N = \{N_i\}$), it is natural to expect that the appropriately chosen set of sensitivity data should be adequate to characterize both charge transfer (CT) and polarization (P) dominated processes, which can be related to the controlling dN and dv perturbations, respectively.

When considering the behaviour of a single molecule or a family of chemically similar molecules, in a given type of a reaction, e.g., during electrophilic, nucleophilic, or radical attack by a small agent, various *single-reactant reactivity concepts* have been successful [3, 7, 8]. They are based upon the underlying notion of an inherent chemical reactivity of a molecule or the relative reactivity of its parts, with respect to a fixed *reaction stimulus*, defined by the character of the other reactant that is assumed to be constant. This notion, implying that the way a molecule reacts is somehow predetermined by its own structure, is clearly very approximate and relatively crude. Obviously, a more subtle *two-reactant description* is required to probe various arrangements of two large reactants, in which the relevant reactivity criteria include the charge sensitivities of both reaction partners, that account for their mutual interaction at any given stage of their approach. It should be emphasized that the sensitivity data are parametrically dependent upon the geometrical structure of the whole reactive system. Therefore, one also has to examine relaxation of the geometry due to the coupling between the geometrical (nuclear position) and electronic (electron population) degrees of freedom of the molecule.

The sensitivity criteria proposed in the litearture fall mainly into the single-molecule criteria [3, 7, 8], although some preliminary analysis of the two-reactant concepts have also been reported [7, 8]. The molecular responses to the (dN, dv)-perturbations are determined to a large extent by the properties of the frontier molecular orbitals, i.e., the *lowest unoccupied* (LUMO) and the *highest occupied* (HOMO) orbitals of both reactants. Therefore, many sensitivity criteria, e.g., FF indices, are closely related to frontier orbital theory [3, 11].

Since the interaction energy is a truly two-reactant concept, all criteria directly linked to it should also be classified as belonging to that category. We would also like to stress that all changes in the sensitivity characteristics of reactants due to the presence of the other reactant, e.g., polarization or relaxation contributions, also have two-reactant character.

Consider a general donor-acceptor system, $M = A - - - B$, consisting of an acidic (A) and a basic (B) reactants. The main one-reactant reactivity criterion, used to predict the system's preferences towards localized attack by small electrophilic, nucleophilic, or radical agents, is based upon the FF indices of the constituent atoms: $f_i = \partial N_i / \partial N$. The reactivity trends follow immediately from the criterion of maximum (minimum) FF index in electrophilic (nucleophilic) attack, that locates the atomic site exhibiting the strongest local basic (acidic) character. This is because the sign of the FF index of a local site in a molecular system directly reflects the site donor/acceptor behaviour in any given type of the global charge displacement, dN. For an electrophilic attack, a molecule acts as a *global base* ($dN < 0$) so a negative FF index implies that the local site acts as a *local acid*, whereas a positive index identifies the *local basic site*. The reverse interpretation applies to *nucleophilic attack*, when the system acts as a *global acid* ($dN > 0$) : a negative FF value now corresponds to the *local basic site*, while a positive FF index identifies the *local acidic site* in the molecule being attacked. Thus, the nucleophilic (B) reactant prefers the site in a molecule (A) with the lowest (preferably negative) FF index, since then the $B \rightarrow A$ coordination is locally enhanced. Similarly, the electrophilic (A) reactant prefers the site in a molecule (B) with the highest (positive) FF value, since then the $A \leftarrow B$ bond is locally strengthened.

The FF data, representing the "renormalized" softness information, carry the same diagnostic potential as the local softnesses. Therefore the latter can also serve as important measures of chemical reactivity, e.g., in metals. As shown by Yang and Parr [3, 12] in their ensemble interpretation, the local softnesses also reflect local fluctuations in the electronic density. This has direct implications for catalytic activity on transition metals, for which such fluctuations in charge have been shown to be very important [13].

It is of interest in the theory of chemical reactivity to examine both open and closed reactive systems M; the former do not preserve N, and are thus free to exchange electrons with a hypothetical reservoir, whereas N remains fixed in the latter, where the basic reactant donates electrons to the acidic reactant without any moderating influence of the M environment. Therefore, the CSA has to define the corresponding quantities for such *external* and *internal CT processes*. The relevant FF indices partition the amount of an external or internal CT into regional or collective contributions. They also directly reflect the corresponding fractions of the CT-energy.

When investigating two large reactants, an adequate reactivity criterion should explicitly take into account the effects of their mutual interaction. Within CSA, such charge couplings are realistically modelled by the rigid hardness matrix, $\eta = \{\partial\mu_i/\partial N_j\}$, where $\mu_i = \partial E/\partial N_i$ is the chemical potential of atom i and E is the energy of the system. As has been shown, [9, 14–16], this matrix is well represented by the corresponding valence-shell electron repulsion integrals, $\eta_{i,j} \approx \gamma_{i,j}$ from the Pariser, Ohno, and related interpolation formulas [17–19], for the actual valence states of *atoms-in-molecules* (AIM). These hardness (interaction) parameters can be considered to be the canonical input data in modelling the associated softness and Fukui function (response) properties. Together they provide a basis for a variety of reactivity concepts. In this article we shall emphasize quantities, electron population reference frames, and the associated charge sensitivities, which take into account the reactant interaction explicitly, and which identify the most important charge reorganization channels in chemical reactions. Another purpose of this new development is to provide a hierarchy of reactivity information, which separates the responses of the reaction region from those of its environment.

In a general reactive system $M = A$ - - - B we assume that all AIM of A are grouped as the first n atoms in η, thus generating the block structure of the AIM's rigid hardness matrix:

$$\eta = \begin{pmatrix} \eta^{A,A} & \eta^{A,B} \\ \eta^{B,A} & \eta^{B,B} \end{pmatrix} \tag{1}$$

The presence of the other reactant modifies η due to its relaxational influence, $\delta\eta(M)$, thus generating the relaxed hardness matrix:

$$\eta^{rel} = \eta + \delta\eta(M) = \begin{pmatrix} \eta^{M}_{A,A} & \eta^{M}_{A,B} \\ \eta^{M}_{B,A} & \eta^{M}_{B,B} \end{pmatrix} \tag{1a}$$

Both matrices can be used as bases for defining two-reactant reactivity concepts. Since $\delta\eta(M) = \delta\eta(\eta)$ [8, 9, 20], such approaches use the same canonical (rigid) AIM-hardness matrix of Eq. (1).

The CSA can be formulated at alternative levels of resolution. Although most of the theoretical development of this work will be presented at the AIM level, that is sufficient for most chemical purposes, several vital formulas related to other resolutions will also be summarized. Each fragment of a molecule has a distinct environment, which moderates the fragment responses to external perturbations. We shall examine in some detail these relaxational (response to dN or dN) and polarizational (response to dv) contributions to charge sensitivities. The *in situ* quantities of the reactants in the A - - - B system, defined by the corresponding derivatives with respect to the amount of N-constrained, internal CT, will also be discussed. Finally, specific concepts for reactive systems, involving collective modes of charge displacements will be defined. The main purpose of this review is to survey reactivity criteria defined within the CSA, which have been developed at the Jagiellonian University, with particular emphasis placed upon the quantities and concepts which can be used to predict activity and selectivity trends as well as the preferred arrangements of very large reactants, e.g., large adsorbate and substrate in the chemisorption systems of heterogeneous catalysis.

2 Molecular Charge Sensitivity Analysis (CSA)

One can define various derivative properties related to the local, regional (molecular fragment) or global descriptions, in which the relevant local density or electron population quantities are treated as independent state variables. Depending upon the resolution, specified by a given partitioning of the system in the physical or function spaces, one considers the following charge-distribution variables: $\rho(r)$ (*local*, L-*resolution*), the electron populations of constituent AIM, $N = (N_1, N_2, ..., N_m)$ or AIM-net charges (AIM-*resolution*), the overall populations attributed to larger molecular fragments (group of AIM), $N_G = (N_X, N_Y, ..., N_Z)$ (*group*, G-*resolution*), or finally, the global number of electrons, $N = \int\rho(r)\,dr = \sum_i N_i = \sum_X N_X$ (*global*, g-resolution). Of course, various intermediate levels, e.g., those defined in the function space spanned by the molecular orbitals (MO-resolution), can also be envisaged, with the corresponding MO occupation numbers as variables, $n = (n_1, ..., n_s)$. The system energy is considered as the function (functional) of these electron population (density) variables: $E^L[\rho]$, $E^{AIM}(\mathbf{N})$, $E^G(\mathbf{N_G})$, and $E^g(N)$. The energy populational gradient defines the chemical potential variables, while the populational (density) hessian represents the hardness matrix (kernel), for a given resolution.

Each resolution implies its own case of the intra-system equilibrium. For example, the L-resolution corresponds to a totally constrained ("frozen") electron distribution, with all local, infinitesimal volume elements considered as being mutually closed, while the g-resolution represents the opposite extreme of a totally relaxed electron distribution, with all local volume elements regarded as mutually open. Similarly, the levels of intermediate resolutions apply to cases of partially constrained equilibrium, in which all fragments defining the partitioning of a molecular system M:

$$M: (r|r'|\ldots)^L, (i|j|\ldots)^{AIM}, (X|Y|\ldots)^G$$

$$(r\mathbin{\vdots}r'\mathbin{\vdots}\ldots)^g = (i\mathbin{\vdots}j\mathbin{\vdots}\ldots)^g = (X\mathbin{\vdots}Y\mathbin{\vdots}\ldots)^g \tag{2}$$

are in their respective internal equilibrium states. As indicated above in the g-resolution all molecular fragments are free to exchange electrons. Thus, for the global equilibrium state of M, the chemical potential (electronegativity) is equalized throughout the whole electron distribution:

$$\delta E^L[\rho]/\delta\rho(r) \equiv \mu(r) = \mu(r') = \ldots$$

$$= \partial E^{AIM}(\mathbf{N})/\partial N_i \equiv \mu_i = \mu_j = \ldots \tag{3}$$

$$= \partial E^G(\mathbf{N_G})/\partial N_X \equiv \mu_X = \mu_Y = \ldots$$

$$= \partial E^g(N)/\partial N \equiv \mu$$

2.1 Hardness and Softness Quantities

2.1.1 Local Resolution

We begin with a survey of the hardness and softness quantities in the local resolution [3, 21, 22]. In this description, the equilibrium (ground-state) density satisfies the Hohenberg–Kohn (HK) variational principle:

$$\frac{\delta E^L[\rho]}{\delta\rho(r)} = \mu = v(r) + \frac{\delta F[\rho]}{\delta\rho(r)} \equiv \mu(r) \tag{4}$$

where $F[\rho]$ is the universal HK density functional generating the sum of the electronic kinetic and repulsion energies. Differentiating this equation with respect to the density gives the hardness kernel in terms of the modified potential $u(r) = v(r) - \mu$:

$$\eta(r', r) = \frac{\delta^2 F[\rho]}{\delta\rho(r')\delta\rho(r)} = -\frac{\delta u(r)}{\delta\rho(r')} \tag{5}$$

The softness kernel is the inverse of the above hardness kernel:

$$\int \eta(r, r') \sigma(r', r'') \, dr' = \delta(r - r'')$$ (6)

$$\sigma(r', r) = -\frac{\delta \rho(r)}{\delta u(r')}$$ (7)

The local softness:

$$s(r) \equiv \left(\frac{\partial \rho(r)}{\partial \mu}\right)_v = \int \frac{\partial u(r')}{\partial \mu} \frac{\delta \rho(r)}{\delta u(r')} \, dr' = \int \sigma(r', r) \, dr'$$ (8)

defines the global softness:

$$S = \partial N/\partial \mu = \int s(r) \, dr = \eta^{-1}$$ (9)

It follows from the Euler Eq. 4 that $\rho = \rho[u]$; therefore:

$$d\rho(r) = \int du(r') \frac{\delta \rho(r)}{\delta u(r')} \, dr' = -\int [dv(r') - d\mu] \, \sigma(r', r) \, dr'$$ (10)

Also, since $\mu = \mu[N, v]$,

$$d\mu = \eta \, dN + \int f(r) dv(r) \, dr$$ (11)

where, by the Maxwell relation:

$$f(r) = \frac{\partial}{\partial N} \frac{\delta E[N, v]}{\delta v(r)} = \left(\frac{\delta \mu}{\delta v(r)}\right)_N = \left(\frac{\partial \rho(r)}{\partial N}\right)_v$$ (12)

These equations give:

$$d\rho(r) = s(r) \eta \, dN + \int dv(r') [-\sigma(r', r) + f(r') s(r)] \, dr'$$ (13)

On the other hand $\rho = \rho[N, v]$,

$$d\rho(r) = f(r) dN + \int dv(r') \beta(r', r) \, dr'$$ (14)

where the linear response function is:

$$\beta(r', r) = \left(\frac{\delta \rho(r)}{\delta v(r')}\right)_N$$ (15)

A comparison between Eqs. (13) and (14) shows that [3, 22]:

$$\beta(r', r) = -\sigma(r', r) + f(r') s(r)$$ (16)

$$f(r) = s(r)/S$$ (17)

This interpretation of β shows that the polarization changes in the electron distribution (responses to the external potential displacements) can be determined from the external softness properties calculated for the fixed nuclear geometry (external potential). This very property is used in determining the mapping relations between the modes of the electron populations and the nuclear positions (see Sect. 2.3).

In the local resolution one derives the following combination formulas for the global chemical potential and hardness:

$$\mu = \left(\frac{\partial E^g}{\partial N}\right)_v = \int \left(\frac{\partial \rho(\mathbf{r})}{\partial N}\right)_v \left(\frac{\delta E^L}{\delta \rho(\mathbf{r})}\right)_v d\mathbf{r} = \int f(\mathbf{r})\mu(\mathbf{r})d\mathbf{r} \tag{18}$$

$$\eta = \left(\frac{\partial^2 E^g}{\partial N^2}\right)_v = \left(\frac{\partial \mu}{\partial N}\right)_v = \iint \left(\frac{\partial \rho(\mathbf{r})}{\partial N}\right)_v \left(\frac{\delta^2 E^L}{\delta \rho(\mathbf{r})\delta \rho(\mathbf{r}')}\right)_v \left(\frac{\partial \rho(\mathbf{r}')}{\partial N}\right)_v d\mathbf{r}d\mathbf{r}' \tag{19}$$

$$= \int f(\mathbf{r})\eta(\mathbf{r},\mathbf{r}')f(\mathbf{r}')d\mathbf{r}d\mathbf{r}'$$

In the global equilibrium state (see the Euler Eq. 4) one obtains the local hardness equalization at the global hardness level:

$$\eta(\mathbf{r}) = \left(\frac{\partial \mu(\mathbf{r})}{\partial N}\right)_v = -\left(\frac{\partial u(\mathbf{r})}{\partial N}\right)_v = -\int \left(\frac{\partial \rho(\mathbf{r}')}{\partial N}\right)_v \left(\frac{\delta u(\mathbf{r})}{\delta \rho(\mathbf{r}')}\right)_v d\mathbf{r}' \tag{20}$$

$$= \int f(\mathbf{r}')\eta(\mathbf{r}',\mathbf{r})d\mathbf{r}' = \left(\frac{\partial \mu}{\partial N}\right)_v = \eta$$

It should be observed that the last two equations are consistent with each other, since $\int f(\mathbf{r})d\mathbf{r} = 1$. One also notes that the local softness and hardness satisfy the following reciprocity relation:

$$\int s(\mathbf{r})\eta(\mathbf{r})d\mathbf{r} = S\eta \int f(\mathbf{r})d\mathbf{r} = 1 \tag{21}$$

The local hardness $\eta(\mathbf{r})$ can be interpreted as the mixed derivative of the energy, comprising differentiation with respect to the global number of electrons and electron density,

$$\eta(\mathbf{r}) = \left(\frac{\partial}{\partial N}\frac{\delta E^L}{\delta \rho(\mathbf{r})}\right)_v = \left(\frac{\partial \mu(\mathbf{r})}{\partial N}\right)_v \tag{22}$$

$$= \left(\frac{\delta}{\delta \rho(\mathbf{r})}\frac{\partial E^g}{\partial N}\right)_v = \left(\frac{\delta \mu}{\delta \rho(\mathbf{r})}\right)_v$$

In Eq. (20) we have used the first part of the above Maxwell relation. Clearly, the same result, equalization of the local hardness, follows from the second part of this identity.

One can interpret the quantity of Eq. (20) as an average over $\mathbf{r}'$ with $f(\mathbf{r}')$ providing the "weighting" factor; it should be stressed, however, that $f(\mathbf{r}')$ exhibits both negative and positive values. Clearly, the derivative of Eq. (20)

does not distinguish between local sites in a molecule, so that it cannot be used as a measure of local differences in hardness of the equilibrium electron distribution.

2.1.2 Atoms-in-Molecules Discretization

Let us start with a brief comment on the nature of the AIM discretization. It can be viewed as resulting from the phenomenological energy expression:

$$E[N, v] = E^{AIM}[N, v] = \sum_i^m N_i v_i + F^{AIM}(N) \tag{23}$$

representing the AIM equivalent of the familiar energy density functional $E^L[\rho] = E_v[\rho] = \int \rho v d\mathbf{r} + F[\rho]$ [1, 2]. Equation (23) introduces the effective potential vector $v = (v_1, ..., v_m)$, where v_i is the electron-nuclei attraction energy per electron on atom i. The AIM chemical potential-equalization equations can now be interpreted as the Euler equations resulting from the variational principle, $\min_N \{E^{AIM}(N, v) - \mu[N(N) - N^\circ]\}$:

$$\frac{\partial E^{AIM}(N, v)}{\partial N_i} = \mu_i = v_i + \frac{\partial F^{AIM}}{\partial N_i} = \mu \tag{24}$$

Continuing this analogy with the local description, one can introduce the modified AIM external potential parameters $\{u_i \equiv v_i - \mu\} = u$, which uniquely define the equilibrium AIM electron populations: $N = N(u)$. In terms of u (see Eqs. (5), (7), and (12):

$$\eta = \frac{\partial^2 F^{AIM}}{\partial N \partial N} = -\frac{\partial u}{\partial N}, \quad \sigma = -\frac{\partial N}{\partial u}, \quad f = \left(\frac{\partial \mu}{\partial v}\right)_N = \left(\frac{\partial N}{\partial N}\right)_v. \tag{25}$$

The AIM softnesses are: $S \equiv (\partial N/\partial \mu)_v = (\partial N/\partial \mu)_v(\partial N/\partial N)_v = Sf$. We can now rederive the Berkowitz and Parr relations of Eqs. (13) and (14) in the AIM resolution:

$$dN(u) = du\frac{\partial N}{\partial u} = -[dv - d\mu \mathbf{1}]\sigma = \eta dNS + dv[-\sigma + f^\dagger Sf]$$

$$= dN[N, v] = dNf + dv\beta \tag{26}$$

which give the AIM equivalents of Eqs. (16) and (17); in Eq. (26) the linear response matrix $\beta \equiv (\partial N/\partial v)_N$.

Consider a given molecular system consisting of m atoms. In what follows we adopt the AIM resolution to define the canonical AIM chemical potentials (electron population gradient), $\mu = \partial E/\partial N = (\mu_1, \mu_2, ..., \mu_m)$, and the corresponding AIM hardness matrix (electron population hessian): $\eta = \partial^2 E/\partial N \partial N = \partial \mu/\partial N = \{\eta_{i,j}\}$; here all differentiations are carried out for the fixed external potential v. This canonical charge-sensitivity information will be used to generate a variety of system charge sensitivities (CS) that probe the responses of the system to various populational perturbations at constant v.

Let us first examine the most important combination problem of determining CS of the system as a whole; we call them global or resultant quantities, since they determine the global equilibrium in the system under consideration. The complementary canonical compliant information is obtained by inverting $\boldsymbol{\eta}$,

$$\boldsymbol{\sigma} \equiv -\partial N/\partial u = \boldsymbol{\eta}^{-1}. \tag{27}$$

This softness matrix can be used to generate the local (regional) and global softness quantities:

$$S_i \equiv \partial N_i/\partial \mu = \sum_j^M \frac{\partial u_j}{\partial \mu} \frac{\partial N_i}{\partial u_j} = \sum_j^M \sigma_{j,i}$$

$$S_x = \partial N_x/\partial \mu = \sum_j^X S_j \tag{28}$$

$$S = \frac{\partial N}{\partial \mu} = \sum_x^M S_x$$

here $\sum_x^X$ denotes the summation over all atomic fragments x in X. The global softness S is the inverse of the global hardness $\eta = \partial \mu/\partial N = 1/S$. The corresponding FF indices are:

$$f_i = \partial N_i/\partial N = (\partial N_i/\partial \mu)/(\partial N/\partial \mu) = S_i/S$$

$$f_x = \partial N_G/\partial N = S_x/S, \text{ etc.} \tag{29}$$

They provide the "weighting" factors for combining the derivatives with respect to the electron-population variables. For example, for the global chemical potential one obtains:

$$\mu = \sum_i^M \frac{\partial N_i}{\partial N} \frac{\partial E^{\text{AIM}}}{\partial N_i} = \sum_i^M f_i \mu_i \tag{30}$$

In terms of the AIM FF indices the global hardness can be expressed as:

$$\eta = \sum_i^M \sum_j^M \frac{\partial N_i}{\partial N} \frac{\partial^2 E^{\text{AIM}}}{\partial N_i \partial N_j} \frac{\partial N_j}{\partial N} = \sum_i^M \sum_j^M f_i \eta_{i,j} f_j \tag{31}$$

It is also of interest to examine the mixed derivative characterizing the resultant local hardness of an atom i, $\eta_i \equiv \dfrac{\partial^2 E^{\text{AIM}}}{\partial N \partial N_i}$

$$\eta_i = \frac{\partial \mu_i}{\partial N} = \sum_j^M \frac{\partial N_j}{\partial N} \frac{\partial \mu_i}{\partial N_j} = \sum_j^M f_j \eta_{j,i} \tag{32}$$

$$= \frac{\partial \mu}{\partial N_i} = \frac{\partial N}{\partial N_i} \frac{\partial \mu}{\partial N} = \eta$$

which is also equalized at the global hardness level; in Eq. (32) we have omitted the constant v constraint to simplify notation. Notice, that Eqs. (31) and (32) imply $\sum_i^M f_i = 1$, which is indeed satisfied by the FF normalization: $\sum_i^M f_i = \sum_i^M \partial N_i/\partial N = \partial N/\partial N = 1$.

We would like to point out that the general second-order expansion for a change in the system energy due to dN,

$$\Delta E(dN) \simeq dN\mu^\dagger + \frac{1}{2}dN\eta dN^\dagger \tag{33}$$

simplifies for the global equilibrium displacement $dN^{eq}(dN) \equiv dNf$, $f = (f_1, f_2, ..., f_m)$, to the familiar expression for the energy in terms of the net change in the global number of electrons, dN:

$$\Delta E[dN^{eq}(dN)] = dN(f\mu^\dagger) + \frac{1}{2}(dN)^2 f\eta f^\dagger = dN\mu + \frac{1}{2}(dN)^2\eta \tag{34}$$

Let us now consider the effective CS of fragments of M for a partitioning of M into two mutually closed, complementary subsystems, M = (A|B), e.g., reactants in the donor-acceptor systems. Of interest in this case are the elements of the condensed hardness matrix, e.g., $\eta^M_{A,B} \equiv (\partial\mu^M_B/\partial N_A)$. Using the chain rule transformation we obtain:

$$\eta^M_{A,B} = \sum_j^M \frac{\partial N_j}{\partial N_A}\frac{\partial\mu_B}{\partial N_j} \equiv \sum_a^A f^{A,A}_a \eta^M_{a,B} + \sum_b^B f^{A,B}_b \eta^M_{b,B} \tag{35}$$

$$\eta^M_{a,B} = \sum_j^M \frac{\partial N_j}{\partial N_B}\frac{\partial\mu_a}{\partial N_j} \equiv \sum_{a'}^A f^{B,A}_{a'} \eta_{a',a} + \sum_b^B f^{B,B}_b \eta_{b,a}, \quad \text{etc.} \tag{36}$$

Here, the diagonal FF indices sum up to 1, $\sum_a^A f^{A,A}_a = \partial N_A/\partial N_A = 1$, while the off-diagonal (relaxational) FF indices sum to 0, $\sum_b^B f^{A,B}_b = \partial N_B/\partial N_A = 0$.

In a similar way, one can express any derivative involving differentiation with respect to the overall fragment electron population. For example, for the group chemical potential one obtains:

$$\mu^M_A = \left(\frac{\partial E^{AIM}}{\partial N_A}\right)_{N_B} = \sum_j^M \frac{\partial N_j}{\partial N_A}\frac{\partial E^{AIM}}{\partial N_j} = \sum_a^A f^{A,A}_a \mu_a + \sum_b^B f^{A,B}_b \mu_b \tag{37}$$

The same chain-rule transformation produces a system of linear equations for determining the relaxed FF indices that appear in Eqs (35-37):

$$f^{A,A}_a = \left(\frac{\partial N_a}{\partial N_A}\right)_{N_B} = \left(\frac{\partial N_a}{\partial N_A}\right)_{N_B} + \sum_b^B f^{A,B}_b T^{(B|A)}_{b,a} \equiv f^{rgd}_a + \delta f_a(M) \tag{38}$$

where the first term represents the *rigid* FF index, calculated for the frozen electron distribution in B, and the second term accounts for the electron relaxation in B measured by the relaxation matrix:

$$T^{(B|A)}_{b,a} = \left(\frac{\partial N_a}{\partial N_b}\right)_{N_A} \tag{39}$$

As shown elsewhere [9] an explicit expression for this matrix element immediately follows from considering changes in the AIM chemical potentials in A, due to dN_b, $\Delta\mu_A(dN_b)$, which are responsible for the subsequent relaxational flows in A, $dN_A(dN_b)$. Namely, in the quadratic approximation a given populational perturbation $dN_b(dN_{b'\neq b} = 0)$ shifts the AIM chemical potentials in A by $\{\Delta\mu_a(dN_b) = dN_b\eta_{b,a}\}$. Thus, the displacements of the local chemical potentials relative to the new equilibrium level, $d\mu_A(dN_b) = dN_b\eta_{b,A}$, are:

$$d\mu_A(dN_b) = \{\Delta\mu_a(dN_b) - d\mu_A(dN_b) = dN_b(\eta_{b,a} - \eta_{b,A})\} \tag{40}$$

The populational displacements in A, relative to the new equilibrium AIM electron populations, $dN_A(dN_b)$, the negative values of the spontaneous relaxational flows $\delta N_A(dN_b)$ that restore the equilibrium, are related to $d\mu_A(dN_b)$ by the matrix electronegativity-equalization equations:

$$d\mu_A = -\delta N_A\eta^{A,A} = dN_A\eta^{A,A} \text{ or } \delta N_A = -dN_A = -d\mu_A\sigma^A \tag{41}$$

where $\eta^{A,A}$ is the diagonal (A,A)-block of η, and $\sigma^A = (\eta^{A,A})^{-1}$. Using Eqs. (40) and (41) gives the desired expression:

$$T_{b,a}^{(B|A)} = \eta_{b,A} S_a^A - \sum_{a'}^A \eta_{b,a'} \sigma_{a',a}^A \tag{42}$$

where $S_a^A = \sum_{a'}^A \sigma_{a',a}^A$.

The equation defining the relaxed off-diagonal FF index is:

$$f_b^{A,B} = \left(\frac{\partial N_b}{\partial N_A}\right)_{N_B} = \sum_a^A \left(\frac{\partial N_a}{\partial N_A}\right)_{N_B}\left(\frac{\partial N_b}{\partial N_a}\right)_{N_B} = \sum_a^A f_a^{A,A} T_{a,b}^{(A|B)} \tag{43}$$

The relaxational matrix also defines the corrections to the relaxed hardness matrix (see Sect. 3.3).

The rigid FF indices of Eq. (38) characterize the uncoupled (non-interacting) molecular fragments. They can be obtained from the non-interacting fragment softness σ^X, $X = (A, B)$, and the associated local and global softnesses. The rigid condensed hardness matrix elements are obtained from Eq. (35) using the rigid FF indices of the reactants:

$$\eta_{A,A}^{rgd} = f_A^{rgd}\eta^{A,A}(f_A^{rgd})^\dagger, \quad \eta_{A,B}^{rgd} = f_A^{rgd}\eta^{A,B}(f_B^{rgd})^\dagger \tag{44}$$

Similarly, one approaches the environmental effects in a general partitioning, $M = (X|Y|Z|...)$, involving more than two mutually closed subsytems (Sect. 3.1). For example, in a catalytic reaction involving two species A and B adsorbed on a crystal C, one explicitly considers the effective charge couplings in $M = (A|B|C)$, characterized by the relevant 3×3 condensed hardness matrix. For such general partitionings into more than two fragments, the off-diagonal blocks, say $\eta^{A,B}$, are also modified, due to the charge relaxation of remaining fragements, by $\delta\eta^{A,B}(M)$.

2.1.3 Relative Internal Development

The preceding section has been devoted to derivative properties involving a net change in the global number of electrons, $dN \neq 0$, i.e., an exchange of electrons between M and a hypothetical external particle reservoir (the sytsem environment). As such, these quantities can be called *external*. However, it is also of interest to examine the *internal* hardness/softness type derivatives, corresponding to $dN = 0$, i.e., the intra system polarization [7, 23]. A typical example is the charge adjustment due to a given displacement of nuclei, which changes the external potential $v(r)$ at constant N.

One way to incorporate the $dN = 0$ constraint is to consider the electron population of one atom, say the m-th, N_m, as dependent on the electron populations of the remaining AIM $N' = (N_1, ..., N_{m-1}: N_m = N_m(N')$ [7, 23]. The closed-system energy in the AIM resolution can therefore be expressed in this relative formulation as a function of the $(m - 1)$ independent variables N':

$$E^{AIM}(N)_N = E^{AIM}[N', N_m(N')] \equiv E_m^{AIM}(N') \tag{45}$$

where, for simplicity, we have dropped the external potential vector in the list of arguments.

The closure relation,

$$dN_m = - \sum_{i=1}^{m-1} dN_i \tag{46}$$

gives the implicit derivatives of N_m with respect to N' $(\partial N_m / \partial N_i)_N = -1$, $i < m$. The corresponding electron-population derivatives of E_m^{AIM}, which take into account this dependence, define the (relative) *internal chemical potentials* of AIM:

$$\mu_i^m \equiv \frac{\partial E^{AIM}}{\partial N_i} + \frac{\partial N_m}{\partial N_i} \frac{\partial E^{AIM}}{\partial N_m} = \mu_i - \mu_m, \quad i < m \tag{47}$$

and the corresponding hardness matrix:

$$H_{j,i}^m = \frac{\partial \mu_i^m}{\partial N_j} + \frac{\partial N_m}{\partial N_j} \frac{\partial \mu_i^m}{\partial N_m} = \eta_{j,i} - \eta_{j,m} - \eta_{m,i} + \eta_{m,m}, \quad (i, j) < m \tag{48}$$

In terms of these quantities, the quadratic energy expression for the closed system becomes:

$$\Delta E_m^{AIM}(N') \cong dN'(\mu^m)^\dagger + \frac{1}{2} dN' H^m (dN')^\dagger \tag{49}$$

where $\mu^m = (\mu_1^m, ..., \mu_{m-1}^m) = \partial E_m^{AIM} / \partial N'$ and the $(m - 1) \times (m - 1)$ matrix $H^m = \{H_{i,j}^m\} = \partial^2 E_m^{AIM} / \partial N' \partial N'$. The electronegativity-equalization equations in this N-constrained approach are:

$$\partial E_m^{AIM} / \partial N' |_{N_e} \equiv \mu_e^m = \mu^m + dN_e' H^m = 0', \tag{50}$$

where N_e' denotes the equilibrium electron populations of the independent atoms; hence the population shifts:

$$dN_e' = N' - N_e' = \mu^m S^m \tag{51}$$

where the *internal* (relative) *softness matrix* $\mathbf{S}^m = \partial \mathbf{N}'/\partial \boldsymbol{\mu}^m = (\mathbf{H}^m)^{-1}$. The $\mathrm{d}N_m$ then follows immediately from the closure constraint of Eq. (46).

Although both $\boldsymbol{\mu}^m$ and $\mathbf{H}^m$ ($\mathbf{S}^m$) are dependent upon the choice of the atom m, the predicted responses $\mathrm{d}\mathbf{N}$ are independent of such a choice; the same is true in the case of other response quantities, e.g., the *internal softness matrix including all m atoms*, $\mathbf{S} \equiv (\partial \mathbf{N}/\partial \boldsymbol{\mu})_N$. This can be explicitly demonstrated by means of the following chain rule transformation:

$$S_{i,j} = \left(\frac{\partial N_j}{\partial \mu_i}\right)_N = \sum_{k=1}^{m-1} \left(\frac{\partial \mu_k^m}{\partial \mu_i}\right)_N \frac{\partial N_j}{\partial \mu_k^m} \equiv \sum_{k=1}^{m-1} \tau_{i,k}^m S_{k,j}^m, \quad (i, j) = 1, \ldots, m \tag{52}$$

where

$$\tau_{i,k}^m = \begin{cases} \delta_{i,k}, & i < m, \\ -1, & i = m, \end{cases} \quad k < m \tag{53}$$

This immediately gives $S_{i,j} = S_{i,j}^m$ for $(i, j) < m$, $S_{m,j} = S_{j,m} = -\sum_{k=1}^{m-1} S_{k,j}^m$ for $j < m$, and $S_{m,m}$ follows from the closure relation of the last column (row) of $\mathbf{S}$, $\sum_{i=1}^{m} S_{i,m} = \sum_{j=1}^{m} S_{m,j} = 0$, $S_{m,m} = \sum_{i=1}^{m-1}\sum_{j=1}^{m-1} S_{i,j}^m$. Thus, the m-independent block of $\mathbf{S}$ is identical with $\mathbf{S}^m$, while the remaining m-dependent elements of $\mathbf{S}$ immediately follow from the relevant closure relations as simple functions of $\mathbf{S}^m$.

Obviously, in symmetrical systems, all atoms which are symmetrically related to the atom m should also be excluded from $\mathbf{N}'$, if the symmetry is to be preserved in the charge displacement process.

The difference between the respective elements of the external (N-unrestricted) and internal (N-restricted) softness matrices, $\sigma_{i,j} = \partial N_j/\partial \mu_i$ and $S_{i,j} = (\partial N_j/\partial \mu_i)_N$, respectively, is due to the external CT component:

$$\sigma_{i,j} = \left(\frac{\partial N_j}{\partial \mu_i}\right)_N + \frac{\partial \mu}{\partial \mu_i}\frac{\partial N}{\partial \mu}\frac{\partial N_j}{\partial N} = S_{i,j} + f_i^M S f_i^M \tag{54}$$

This equation represents the AIM equivalent of the Berkowitz and Parr relation of Eq. (16):

$$\sigma(\mathbf{r}', \mathbf{r}) = -\beta(\mathbf{r}', \mathbf{r}) + f(\mathbf{r}') S f(\mathbf{r}) \tag{55}$$

Thus,

$$\boldsymbol{\beta} = \{\beta_{i,j} \equiv (\partial N_j/\partial v_i)_N = -S_{i,j}\} \tag{56}$$

The basic relations determining CS in the AIM resolution of M as a whole, which were presented in the preceding sections, can be summarized in a simple algorithm organized as a single matrix-inversion step. Assume that the molecular system M was shifted from an initial, global equilibrium ($\mu_1^0 = \mu_2^0 = \ldots = \mu_m^0 = \mu^0$) by an external perturbation $(\mathrm{d}N, \mathrm{d}v)$; as a result the system reaches the displaced equilibrium state corresponding to a new level of the global chemical potential: $\mu_1 = \mu_2 = \ldots = \mu_m = \mu(\mathrm{d}N, \mathrm{d}v)$. In accordance with our previous development, we first determine the displaced AIM chemical potentials, $\boldsymbol{\mu}^+ \equiv \mu^0 \boldsymbol{1} + \mathrm{d}\boldsymbol{v}$, using the resulting chemical potential displacements $\mathrm{d}\boldsymbol{\mu}(\mathrm{d}\boldsymbol{v}) = (\boldsymbol{\mu}^+ - \mu\boldsymbol{1})$ as the forces behind subsequent net changes in the AIM electron populations, due to both $\mathrm{d}N$ and $\mathrm{d}\boldsymbol{v}$, $\boldsymbol{\delta N} = (N_1 - N_1^0, ..., N_m - N_m^0)$. The corresponding electronegativity-equalization equations, that link the initial and final chemical potentials and incorporate the obvious relation $\sum_i^M \mathrm{d}N_i = \mathrm{d}N$, may be written in the following matrix form:

$$(-\mu, \boldsymbol{\delta N})\begin{pmatrix} 0 & \boldsymbol{1} \\ \boldsymbol{1}^\dagger & \boldsymbol{\eta} \end{pmatrix} \equiv \boldsymbol{\mathfrak{R}}\tilde{\boldsymbol{\eta}} = (\mathrm{d}N, -\boldsymbol{\mu}^+) \equiv \boldsymbol{\wp} \tag{57}$$

where $\boldsymbol{\mathfrak{R}}$ and $\boldsymbol{\wp}$ represent the generalized response and perturbation vectors, respectively. One can solve this equation by inverting the generalized hardness matrix $\tilde{\boldsymbol{\eta}}$:

$$(-\mu, \boldsymbol{\delta N}) = \boldsymbol{\wp}\tilde{\boldsymbol{\eta}}^{-1} = (\mathrm{d}N, -\boldsymbol{\mu}^+)\begin{pmatrix} -\eta & \boldsymbol{f} \\ \boldsymbol{f}^\dagger & -\boldsymbol{\beta} \end{pmatrix} \equiv \boldsymbol{\wp}\,\tilde{\boldsymbol{\sigma}} \tag{58}$$

where $\tilde{\boldsymbol{\sigma}}$ denotes the generalized softness matrix. Indeed, the first component of this equation reads: $\mu - \mu^0 \equiv \mathrm{d}\mu(\mathrm{d}N, \mathrm{d}\boldsymbol{v}) = \mathrm{d}N\eta + \mathrm{d}\boldsymbol{v}\boldsymbol{f}^\dagger$, while the second component expresses $\boldsymbol{\delta N}(\mathrm{d}N, \mathrm{d}\boldsymbol{v}) = \mathrm{d}N\,\boldsymbol{f} + \boldsymbol{\mu}^+\boldsymbol{\beta} = \mathrm{d}N\boldsymbol{f} + \mu^0\boldsymbol{1}\boldsymbol{\beta} + \mathrm{d}\boldsymbol{v}\boldsymbol{\beta} = \mathrm{d}N\boldsymbol{f} + \mathrm{d}\boldsymbol{v}\boldsymbol{\beta}$, since $\sum_i^M \beta_{i,j} = (\partial N/\partial v_i)_N = 0$.

2.2 Collective Charge Displacements and Their Charge Sensitivities

The AIM electron-population displacements, $\mathrm{d}N$, are strongly coupled through the off-diagonal hardness matrix elements $\eta_{i,j} \approx \gamma_{i,j}$. Thus, a given displacement $\mathrm{d}N_k$ strongly affects the chemical potentials of all AIM. This representation considers all AIM populational parameters as independent variables, which can be interpreted as projections of the *populational vector* $\boldsymbol{\mathcal{N}} = (N_1 \hat{\boldsymbol{e}}_1 + N_2 \hat{\boldsymbol{e}}_2 + \cdots N_m \hat{\boldsymbol{e}}_m)$ onto the orthogonal system of populational axes associated with the constituent atoms, i.e., the AIM populational basis vectors: $\hat{\boldsymbol{e}}_M = (\hat{\boldsymbol{e}}_1, ..., \hat{\boldsymbol{e}}_m)$. In what follows, the products of vectors represent scalar products.

A more compact description of chemical reactivity can be obtained in representations involving collective charge displacements [7, 9, 14, 24–29]. $\mathrm{d}\boldsymbol{p} = \mathrm{d}N\,\mathbf{T}$, defined by the transformation matrix $\mathbf{T}$; they also represent a degree of mode decoupling. In this section we shall summarize the most important sets of such collective (delocalized) modes of potential interest in the theory of chemical reactivity. We start with the total decoupling, which is obtained in the representation of the hardness matrix eigenvectors, called *populational normal modes* (PNM) [7, 8, 14, 29]. They are analogous to the normal modes of nuclear motions defined in nuclear position space. Next, we consider the (non-orthogonal) set of *minimum energy coordinates* (MEC) [7, 8, 27, 28], defined in the softness representation, again in direct analogy to the compliance formalism of nuclear motions [30, 31].

2.2.1 Populational Normal Modes

The totally decoupled representation involves displacements in populations of PNM,

$$d\boldsymbol{n} = d\boldsymbol{N}\boldsymbol{U}, \quad d\boldsymbol{\mathscr{N}} = \boldsymbol{\bar{U}}_M d\boldsymbol{n}^\dagger \tag{59}$$

defined by the columns of the transformation matrix $\boldsymbol{T} = \boldsymbol{U} \equiv (\boldsymbol{U}_1 | \boldsymbol{U}_2 | ... | \boldsymbol{U}_m)$, which diagonalizes the AIM hardness matrix, i.e., "rotates" the AIM axes $\boldsymbol{\hat{e}}_M$ to the (orthogonal) principal axes of the hardness tensor (see Fig. 1), $\boldsymbol{\bar{U}}_M = \boldsymbol{\hat{e}}_M \boldsymbol{U}$:

$$\boldsymbol{U}^\dagger \boldsymbol{\eta} \boldsymbol{U} = \boldsymbol{h} = \{h_\alpha \delta_{\alpha\beta}\}, \quad \boldsymbol{U}^\dagger \boldsymbol{U} = \boldsymbol{I} \tag{60}$$

where $h_\alpha = \partial^2 E^{PNM}/\partial n_\alpha^2 = \partial \xi_\alpha/\partial n_\alpha \equiv 1/\mathscr{S}_\alpha$ is the principal hardness of mode α, and $\xi_\alpha = \mu \boldsymbol{U}_\alpha$ is the chemical potential of the mode. Obviously, the quadratic form of the second differential of the energy of the system with respect to the electron population variables becomes diagonal in the $\boldsymbol{\bar{U}}_M$ reference frame:

$$d^2 E^{AIM} \equiv \frac{1}{2} d\boldsymbol{N}\boldsymbol{\eta} d\boldsymbol{N}^\dagger = \frac{1}{2}\sum_\alpha^M h_\alpha (dn_\alpha)^2 \equiv d^2 E^{PNM} \tag{61}$$

Also decoupled are the corresponding electronegativity equalization equations (see Eq. (41)):

$$d\xi_\alpha = dn_\alpha h_\alpha \quad \text{or} \quad d\boldsymbol{\xi} = d\boldsymbol{n}\boldsymbol{h} \tag{62}$$

Here $d\boldsymbol{\xi} = (d\xi_1, ..., d\xi_m)$ and $d\boldsymbol{n} = (dn_1, ..., dn_m)$ are the displacements from the corresponding equilibrium (global) values. The global chemical potential and hardness can now be formally expressed in terms of $\boldsymbol{\xi}$,

$$\mu = \frac{\partial \boldsymbol{n}}{\partial N} \frac{\partial E^{PNM}}{\partial \boldsymbol{n}} \equiv \boldsymbol{F}\boldsymbol{\xi}^\dagger \tag{63}$$

$$\eta = \boldsymbol{f}\boldsymbol{\eta}\boldsymbol{f}^\dagger = \sum_\alpha^M F_\alpha^2 h_\alpha \tag{64}$$

where the normal FF indices $F_\alpha = \partial n_\alpha/\partial N = \boldsymbol{f}\boldsymbol{U}_\alpha$.

The phases of the CT-active PNM are fixed by the convention [26]:

$$\varphi_\alpha = \frac{\partial N}{\partial n_\alpha} = \frac{\partial N}{\partial n_\alpha}\frac{\partial N}{\partial N} = \boldsymbol{1}\boldsymbol{U}_\alpha > 0 \tag{65}$$

in analogy with the equivalent AIM space equation: $\partial N/\partial N_i = 1$. We would like to point out, that antisymmetric modes with respect to a symmetry operation of the system in question, have $\varphi_\alpha = 0$, since they are purely polarizational, and have no external CT component.

It should be observed that for the global equilibrium charge distribution, where $\mu = \mu\boldsymbol{1}$,

$$\boldsymbol{\xi}^{eq} = \mu\boldsymbol{1}\boldsymbol{U} = \mu\boldsymbol{\varphi} = \frac{\partial E^g}{\partial N}\frac{\partial N}{\partial \boldsymbol{n}} \tag{66}$$

Consider now the normalization of the FF indices $\boldsymbol{F} = (F_1, ..., F_m)$ of Eq.

(63). Obviously, since the rotation of $\boldsymbol{\acute{e}}_M$ into $\vec{\boldsymbol{U}}_M$ does not preserve

$$\mathrm{d}N = \boldsymbol{1}\,\mathrm{d}\boldsymbol{N}^\dagger \neq \boldsymbol{1}\,\mathrm{d}\boldsymbol{n}^\dagger = \sum_i^M \mathrm{d}N_i \sum_\alpha^M U_{i,\alpha} \tag{67}$$

the normal FF indices do not sum up to 1:

$$\sum_\alpha^M F_\alpha = \sum_\alpha^M \frac{\partial n_\alpha}{\partial N} \neq 1 \tag{68}$$

As a consequence of these two equations neither the equilibrium normal chemical potentials, $\boldsymbol{\xi}^{\mathrm{eq}} = \mu\boldsymbol{\varphi}$, nor the equilibrium mode hardnesses

$$\boldsymbol{H}^{\mathrm{eq}} \equiv \partial\boldsymbol{\xi}/\partial N = \partial\mu/\partial n = \eta\boldsymbol{\varphi} \tag{69}$$

are equalized at the global equilibrium; this may create some interpretational difficulties.

It would be highly desirable to bring the normal-mode description as close as possible to the AIM interpretation, with these two equalization principles being automatically satisfied and the appropriate FF indices providing the real-mode contributions to the unit change in the global number of electrons. In order to reach this goal, one has to redefine the mode-population variables, by projecting the $\mathrm{d}\boldsymbol{n}$ displacements of Eq. (59) onto the pure CT-direction represented by the vector:

$$\boldsymbol{\acute{e}}^N = \sum_i^M \left(\frac{\partial N}{\partial N_i}\right)\boldsymbol{\acute{e}}_i = \boldsymbol{1}\,\boldsymbol{\acute{e}}_M^\dagger \tag{70}$$

which determines the shift in the global number of electrons, $\mathrm{d}N = \boldsymbol{\acute{e}}^N\mathrm{d}\vec{\mathscr{N}} = \boldsymbol{1}\,\mathrm{d}\boldsymbol{N}^\dagger$. This procedure is also illustrated in Fig. 1. The CT-projected shifts in mode populations,

$$\mathrm{d}\bar{n}_\alpha = \mathrm{d}n_\alpha\frac{\partial N}{\partial n_\alpha} = \mathrm{d}n_\alpha\varphi_\alpha, \quad \alpha = 1,\dots,m \tag{71}$$

are phase-independent and correspond to the non-unitary transformation of the AIM population displacements:

$$\mathrm{d}\bar{\boldsymbol{n}} = \mathrm{d}\boldsymbol{N}\bar{\boldsymbol{U}}; \quad \bar{\boldsymbol{U}} = (\varphi_1\,\boldsymbol{U}_1|\varphi_2\,\boldsymbol{U}_2|\dots|\varphi_m\,\boldsymbol{U}_m) \tag{72}$$

The inverse transformation is:

$$\mathrm{d}\boldsymbol{N} = \mathrm{d}\bar{\boldsymbol{n}}\bar{\boldsymbol{U}}^{-1}; \quad (\bar{\boldsymbol{U}}^{-1})^\dagger = (\varphi_1^{-1}\,\boldsymbol{U}_1|\varphi_2^{-1}\,\boldsymbol{U}_2|\dots|\varphi_m^{-1}\,\boldsymbol{U}_m) \tag{73}$$

It can be easily verified that:

$$\sum_\alpha^M \mathrm{d}\bar{n}_\alpha = \mathrm{d}N \tag{74}$$

so that the associated CT-projected FF indices,

$$\boldsymbol{w} = \frac{\partial\bar{\boldsymbol{n}}}{\partial N} = (\varphi_1\,F_1, \varphi_2\,F_2, \dots, \varphi_m F_m) \tag{75}$$

exhibit the desired "normalization":

$$\sum_{\alpha}^{M} w_{\alpha} = \frac{\partial N}{\partial N} = 1 \tag{76}$$

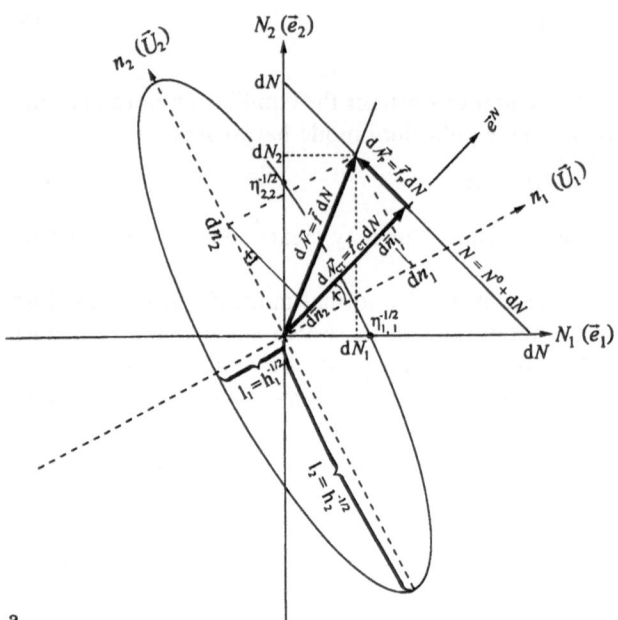

a

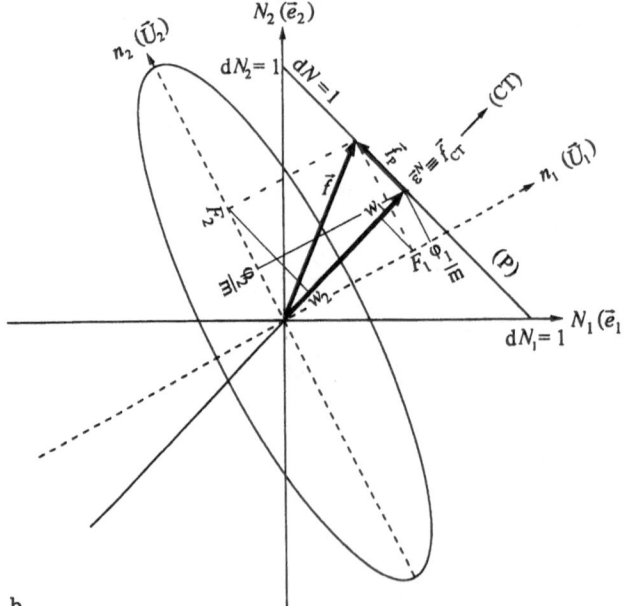

b

Moreover, since

$$\bar{\varphi} = \frac{\partial N}{\partial \bar{n}} = 1 \tag{77}$$

one obtains the desired AIM-like equalization of the new mode chemical potentials:

$$\bar{\xi}^{eq} = \frac{\partial \bar{E}^{PNM}(\bar{n})}{\partial \bar{n}} = \frac{\partial N}{\partial \bar{n}} \frac{\partial E^g}{\partial N} = \mu \mathbf{1} \tag{78}$$

and the modified mode hardnesses:

$$\bar{H}^{eq} = \frac{\partial \bar{\xi}}{\partial N} = \frac{\partial \mu}{\partial \bar{n}} = \frac{\partial N}{\partial \bar{n}} \frac{\partial \mu}{\partial N} = \eta \mathbf{1} \tag{79}$$

We again emphasize that for a non-equilibrium charge distribution, e.g., from independent SCF calculations, $\bar{\xi} = \mu \bar{U}$, may not be equalized. It should be observed that the AIM components of w_α indices,

$$w_\alpha = F_\alpha \sum_i^M \frac{\partial N_i}{\partial n_\alpha} = \sum_i^M F_\alpha U_{i,\alpha} \equiv \sum_i w_{i,\alpha} \tag{80}$$

also provide a resolution of the AIM FF indices into the PNM contributions:

$$f_i = \frac{\partial N_i}{\partial N} = \frac{\partial n}{\partial N} \frac{\partial N_i}{\partial n} = \sum_\alpha^M F_\alpha U_{i,\alpha} = \sum_\alpha^M w_{i,\alpha} \tag{81}$$

In terms of the CT-projected population variables, the hardness tensor is given by the diagonal matrix:

$$\bar{h} \equiv \frac{\partial^2 \bar{E}^{PNM}(\bar{n})}{\partial \bar{n} \partial \bar{n}} = \frac{\partial \bar{\xi}}{\partial \bar{n}} = \left\{ \frac{1}{\varphi_\alpha^2} h_\alpha \delta_{\alpha\beta} \equiv \bar{h}_\alpha \delta_{\alpha, \beta} \right\} \tag{82}$$

the corresponding expression for the PNM softness matrix is:

$$\bar{\mathscr{S}} = \bar{h}^{-1} = \frac{\partial \bar{n}}{\partial \bar{\xi}} = \left\{ \varphi_\alpha^2 \mathscr{S}_\alpha \delta_{\alpha\beta} \equiv \bar{\mathscr{S}}_\alpha \delta_{\alpha\beta} \right\} \tag{83}$$

Fig. 1. The PNM (principal axes), $\bar{U}_M$, in the system with two AIM populational degrees of freedom, $\hat{e}_M$, the associated charge sensitivities, and a separation of the pure CT and P components, $d\bar{\mathcal{N}}_{CT}$ and $d\bar{\mathcal{N}}_P$, of the equilibrium charge displacement $d\bar{\mathcal{N}} = \bar{f} dN$, along the FF vector $\bar{f} = \hat{e}_M f^\dagger = \bar{U}_M F^\dagger$. **Panel (a)** refers to a general displacement $dN > 0$, while **Panel (b)** corresponds to $dN = 1$. The hardness tensor ellipse with semiaxes of length $\{l_\alpha\}$, is defined by the quadratic surface: $dN \eta dN^\dagger = \sum_\alpha^M h_\alpha n_\alpha^2 \equiv \sum_\alpha^M n_\alpha^2 / l_\alpha^2 = 1$. As explicitly shown in **Panel (a)**, the shifts of the modified PNM population variables, $d\bar{n}$, are obtained by projecting the relevant PNM variables, dn, onto the pure-CT direction $\hat{e}^N = \mathbf{1} \hat{e}_M^\dagger$. One similarly obtains a resolution of $d\bar{\mathcal{N}}$ into $d\bar{\mathcal{N}}_{CT} = d\bar{\mathcal{N}} \cdot \hat{e}^N$ and $d\bar{\mathcal{N}}_P = d\bar{\mathcal{N}} - d\bar{\mathcal{N}}_{CT}$ [**Panel (a)**], and of $\bar{f}$ into $f_{CT} = \bar{f} \cdot \hat{e}^N$ and $\bar{f}_P = \bar{f} - \bar{f}_{CT}$ [**Panel (b)**]

where $\mathscr{S}_\alpha \equiv h_\alpha^{-1}$. Therefore, the purely polarizational ($\varphi_\alpha = 0$) modes (internal redistribution channels) have infinite CT-projected-mode hardness, $\bar{h}_\alpha \equiv \infty$ ($\mathscr{S}_\alpha = 0$).

Consider now a unit displacement in the global number of electrons $dN = 1$ (see Fig. (1b)). The equilibrium AIM electron population displacements are then given by $\vec{f} = \hat{e}_M \, f^\dagger$; together with its CT component, $\vec{f}_{CT} = (1/m) \, \hat{e}_M^\dagger \equiv \bar{e}^N \hat{e}_M^\dagger$, it defines the CT-induced polarization vector: $\vec{f}_P = \vec{f} - \vec{f}_{CT}$, where $\vec{f}_{CT}$ represents the scaled CT) vector $\hat{e}^N = \partial N/\partial N = \boldsymbol{\varphi}^{AIM} = 1$, "renormalized" to represent a single electron displacement. Obviously, the external CT to/from the system, involving an electron reservoir, will also induce the associated internal polarization (P) in a molecule, so that the two vectors $\vec{f}$ and $\vec{f}_{CT}$ do not coincide, due to a non-vanishing P-component. It is of interest in the theory of chemical reactivity to distinguish between these two components and provide their AIM, PNM and (AIM + PNM) representations, respectively.

When considering the PNM representation, one should remember that the corresponding FF quantities include the overall (P + CT) displacement, while the renormalized $\boldsymbol{\varphi}$-derivatives represent the CT part alone. Thus, in the PNM framework, the $\boldsymbol{F}$ components represent the resultant (P + CT) vector $\vec{f}$; the corresponding $\vec{f}_{CT}$ components are given by the vector $\boldsymbol{\varepsilon}^N \equiv (1/m)$ $(\partial N/\partial n) = (1/m)\boldsymbol{\varphi}$, and the $\vec{f}_P$ components are defined by the vector $(\boldsymbol{F} - \boldsymbol{\varepsilon}^N) = \{F_\alpha - \varphi_\alpha/m\}$ (see Fig. 1b). Of great interest also is the (AIM + PNM)-resolution. As we have observed above, $f_i = \sum_\alpha^M w_{i,\alpha}$; also:

$$\frac{1}{m}\varphi_i^{AIM} = \frac{1}{m}\frac{\partial N}{\partial N_i} = \frac{1}{m}\sum_\alpha^M \frac{\partial N}{\partial n_\alpha}\frac{\partial n_\alpha}{\partial N_i} = \frac{1}{m}\sum_\alpha^M \varphi_\alpha U_{i,\alpha} = \frac{1}{m}\sum_\alpha^M \bar{U}_{i,\alpha}. \quad (84)$$

Therefore, the quantities $\bar{U}_{i,\alpha}/m$ and $(w_{i,\alpha} - \bar{U}_{i,\alpha}/m)$ provide the contributions to the CT and P components, respectively, from the mode α at the position of the i-th AIM, of the unit equilibrium charge displacement of $dN = 1$ in M.

The transformation $\bar{U} = \partial\bar{n}/\partial N$ measures the CT-projected shifts in the mode populations, per unit test displacements in the AIM population N. Thus, all elements of the columns of $\bar{U}$ that represent purely P-modes, must vanish identically ($\varphi_\alpha = 0$). The associated rows of the inverse transformation $\bar{U}^{-1} = \partial N/\partial\bar{n}$, providing similar measures of response in the AIM populations per unit test displacements in the mode populations $\bar{n}$, must therefore be infinite. Indeed, since they may be considered as corresponding to $\bar{n}_\alpha \to 0$, a finite $\Delta\bar{n}_\alpha = 1$ has to correspond to infinite AIM displacements.

In order to illustrate the above normal-mode concepts and quantities, we have generated the relevant charge sensitivities for toluene, using the MNDO [32] AIM charges and the corresponding Ohno interpolated hardness tensor. The results are displayed in Figs. 2 and 3, with modes ordered in accordance with the increasing principal hardness, h_α. The highest value of h_α (lowest value of $\bar{h}_\alpha$), corresponds to the non-selective mode $\alpha = 15$ in which electron populations of all AIM are shifted "in phase" by approximately the same amount, i.e., when all of them simultaneously accept (or lose) a comparable number of electrons; thus, this mode exhibits practically pure CT-character ($w_\alpha \simeq 1$), and

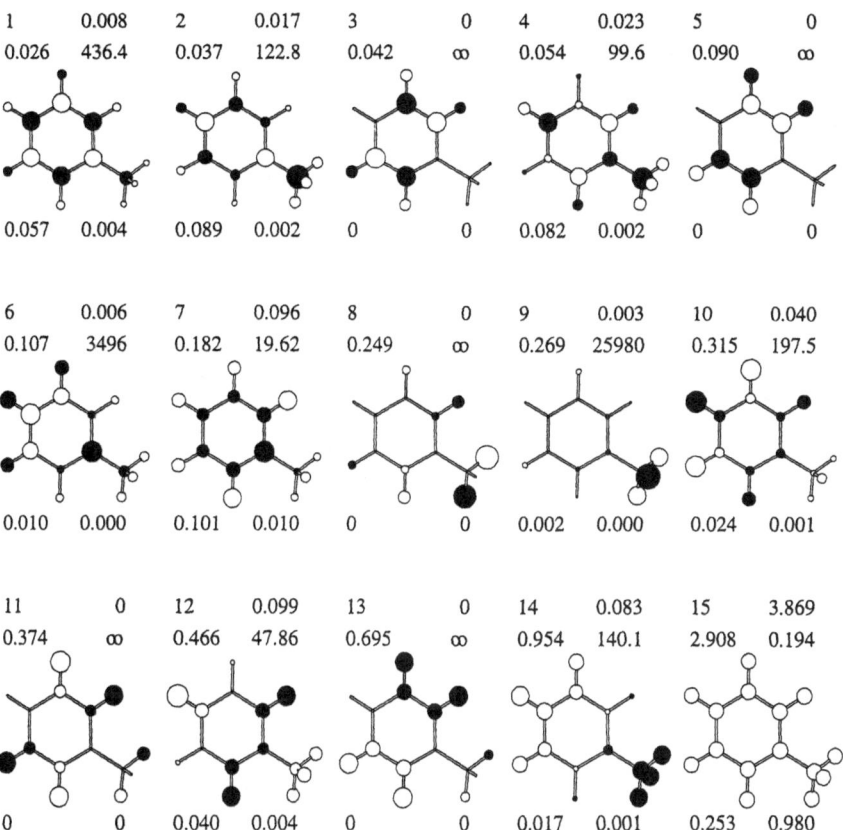

Fig. 2. Schematic diagrams of the PNM in toluene, with the shaded areas denoting negative values. In **Panel (a)** the area within a circle on atom i in the mode α is proportional to $|U_{i,\alpha}|$, with all diagrams plotted using the same scaling factor, and the CT-active mode phases determined by the convention $\varphi_\alpha > 0$. The mode ordering is in accordance with increasing values of h_α. The numbers reported (a.u.) are: φ_α (first row), h_α, $\bar{h}_\alpha$ (second row), and F_α, w_α (third row, below the diagram)

zero nodes in the mode contour. The h_α–softest mode, $\alpha = 1$, corresponding to the lowest eigenvalue, exhibits practically pure P-character ($w_\alpha \simeq 0$), and the maximum number of nodes in the mode diagram. One could say roughly, that the h_α–hardest and –softest modes exhibit, respectively, the dominant channels for an *external CT* (to/from the system in question) and the *CT-induced internal polarization*. It should be observed that it is the softest (site-selective) mode, representing the most facile adjustment of the AIM electron distribution, that determines the internal differentiation of the AIM charges, as a result of the charge donated (accepted) through the hardest mode, which is totally non-selective.

Only the ten modes that are symmetric with respect to the symmetry plane of the toluene molecule can be CT-active. The five antisymmetric modes represent

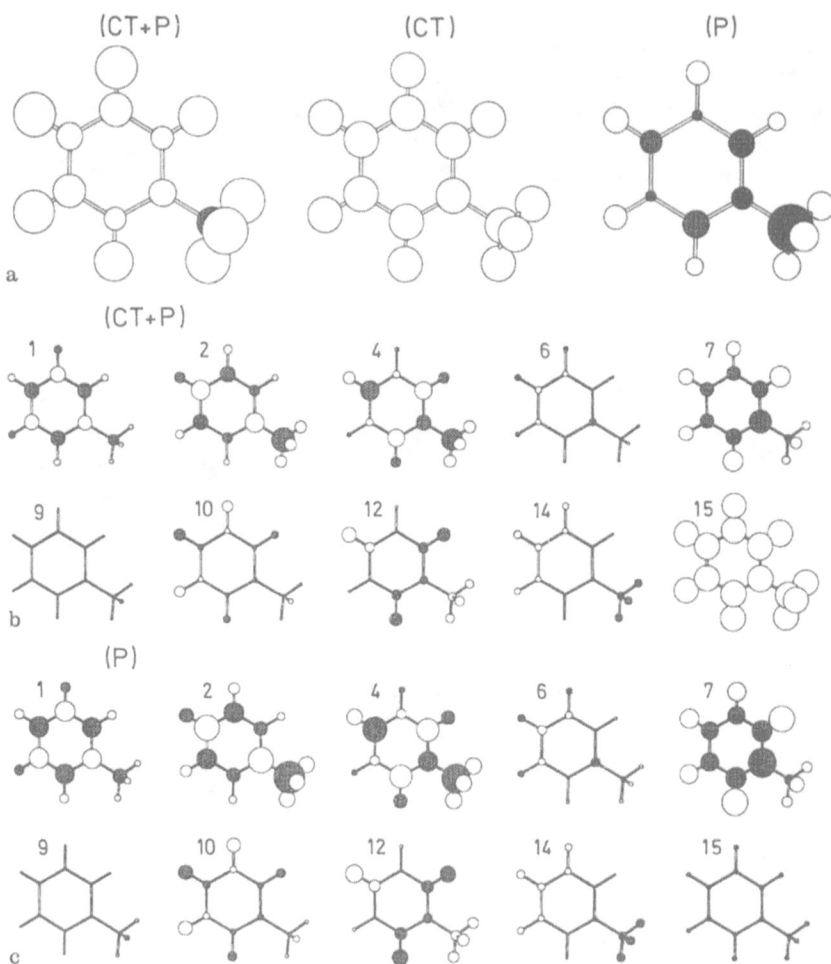

Fig. 3 The AIM (**Panel a**) and AIM + PNM (**Panels b** and **c**) resolutions of the (P + CT), CT, and P population displacements for $dN = 1$, represented by the vectors, $\vec{f} = f \partial \hat{\imath}_M =$ $F \check{U} \hat{\imath}_M$, $\bar{e}^N = e^N \partial \hat{\imath}_M$ and $(\vec{f} - \bar{e}^N)$, respectively. The areas of the circles in three diagrams of panel (a) are proportional to $|f_i|$, $1/m$, and $|f_i - 1/m|$, respectively; in **Panel (b)** and **(c)**, where only the CT-active ($\varphi_\alpha > 0$) modes are shown, the two groups of diagrams represent $w_{i,\alpha}$ and $(w_{i,\alpha} - \check{U}_{i,\alpha}/m)$ quantities, respectively

the purely internal polarization of the system, for constant N. For such modes $\varphi_\alpha = w_\alpha = F_\alpha = \xi_\alpha = 0$. In terms of the modified hardnesses $\bar{h}_\alpha$ and chemical potential $\bar{\xi}_\alpha$, such P-modes have infinite values of $\bar{h}_\alpha$ and by the de l'Hospital rule, $\bar{\xi}_\alpha = \xi_\alpha/\varphi_\alpha = (\partial\xi_\alpha/\partial n_\alpha)/(\partial\varphi_\alpha/\partial n_\alpha)$ are equal to $(+\infty)$ for a stable mode and $(-\infty)$ for an unstable mode, in accordance with the signs of the principal hardnesses in the numerator. Clearly, when more symmetry operations are present in the point group of the system, only the totally symmetric modes can be CT-active.

48

The PNM chemical-potential parameters, $\boldsymbol{\xi}^{eq} = \mu\boldsymbol{\varphi}$, provide yet another natural classification of the normal modes into three groups: (a) *acceptor* PNM ($\xi_\alpha < 0$) , (b) *donor* PNM ($\xi_\alpha > 0$), and (c) P-modes ($\xi_\alpha = 0$). The energetic rationale for this distinction becomes clear when one considers the contributions to the energy of the system, $dE_M^{(1)} = \boldsymbol{\xi}dn(dN)^\dagger = \sum_\alpha^M \xi_\alpha F_\alpha dN$, due to changes in the PNM populations, $dn(dN)$. The donor modes stabilize the system when they decrease their mode populations, which is the case when $F_\alpha > 0$, i.e., when the mode α population changes in phase with that of the system as a whole ($dN < 0$). In the case of the acceptor mode, only its population increase stabilizes the system for $F_\alpha > 0$ when $dN > 0$. It should also be noted that these predictions are reversed for the $F_\alpha < 0$ modes, the populations of which are shifted in the directions opposite to dN. In other words, in an electrophilic attack ($dN < 0$) the CT stabilization energy results from the dominant $F_\alpha > 0$ donor modes and – possibly – a few $F_\alpha < 0$ acceptor modes; the opposite is true in the nucleophilic attack ($dN > 0$), the stabilization energy of which originates from the dominant $F_\alpha > 0$ acceptor modes and the $F_\alpha < 0$ donor modes; clearly, all remaining CT-active modes increase the system energy.

The dominant role of the h_α-hardest mode, $\alpha = 15$, immediately follows from its w_α value and the $\boldsymbol{U}_{15}$ vector coefficients (see Fig. 2). The same conclusion, therefore, must follow from the relevant resolution of the pure CT-vector, given by the $\boldsymbol{U}$ transformation, uniformly scaled by the factor $1/m$. This feature of the $\alpha = 15$ mode is also seen in the (AIM + PNM) resolution of the FF vector, shown in Fig. 3b, where all but last CT-active modes mainly describe the CT-induced polarization (explicitly extracted in Fig. 3c).

The h_α-hierarchy of PNM, used as the ordering criterion, identifies the softest mode as the one with the largest number of nodes in the mode diagram. This is opposite to the familiar relation between the nodal structure and energy of the canonical molecular orbitals (MO), for which the lowest orbital energy corresponds to the most delocalized (the least number of nodes) occupied MO, while the highest orbital energy identifies the antibonding MO with the largest number of nodes. A reference to the $\bar{h}_\alpha$ values in Fig. 2 shows that this modified mode hardness brings the PNM description closer to the above MO analogy. Namely, the CT-dominant mode is now identified by the *lowest* value of $\bar{h}_\alpha$. Moreover, the mode $\alpha = 7$, which dominates the CT-induced polarization in toluene (second largest w_α; compare also the mode $\alpha = 7$ contours of Figs. 2 and 3c, with the AIM resolved polarization part of f in Fig. 3a), appears as the second $\bar{h}_\alpha$-softest mode. Thus the $\bar{h}_\alpha$ values do indeed reflect the PNM participation in the CT-related charge rearrangement, with the $\bar{h}_\alpha$-soft modes participating the most. Fig. 2 also shows that with increasing h_α, larger groups of neighbouring atoms change their electron population in phase, thus representing long range polarization; the h_α-soft modes are seen to be responsible for the close-range polarization between neighbouring atoms.

It follows from Figs. 3b and 3c that the CT-induced polarization mostly involves modes $\alpha = 1, 2, 4, 7, 10,$ and 12, each exhibiting a small CT component measured by the $w_\alpha \neq 0$ indices. We would like to remark at this point that the non-selective mode $\alpha = 15$ is of little interest in the theory of chemical reactivity, since it practically affects all AIM uniformly. The CT-active polarizational modes differentiate net changes on constituent atoms as a result of external CT, and – as such – they influence the reactivity/selectivity trends. In toluene this CT-induced polarization component (last diagram in Fig. 3a) is mainly polarization between the carbons and surrounding hydrogens, with $dN > 0$ (an inflow of electrons, e.g., due to an attack by a nucleophile), partly reversing the $C-H$ polarization in an isolated toluene molecule; the reverse charge rearrangement, enhancing this initial polarization, is predicted for $dN < 0$ (an outflow of electrons, e.g., due to an attack by an electrophile). As also seen in Fig. 3a, the methyl carbon exhibits a negative FF index, thus changing its electron population in the opposite way to that in the system as a whole.

2.2.2 Minimum Energy Coordinates

Following the compliance formalism of nuclear displacements [30, 31], one can define the corresponding concepts of the populational *minimum energy coordinates* (MEC) and the associated compliance constants in the space of the AIM electron populations [7, 8, 27, 28]. They can be defined for both open ($dN \neq 0$) and closed ($dN = 0$) systems; however, it has been shown numerically [7] that the MEC characteristics remain practically the same in both the *external* and *internal* cases. Therefore, in an illustrative application to toluene, we shall report only the open system MEC and its hardness parameters. The MEC concept can also be given the relaxational interpretation, as the minimum-energy path of the electron redistribution in the molecular environment of the primarily displaced atom. We shall call such collective charge displacements in the molecular-fragment resolution the *relaxational coordinates* (REC). Since all these concepts represent the responses of the system to localized charge displacements (test oxidations or reductions) they obviously are of great interest in the theory of chemical reactivity [7, 8, 27, 28].

Let us first define the external MEC in M, consisting of m atoms. Consider the global equilibrium of M in contact with a hypothetical electron reservoir (r): $\mu_0 = \mu \mathbf{1}$ where $\mu = \mu_r$, the chemical potential of r. Let $\mathbf{z} = \mathbf{N} - \mathbf{N}^0 \equiv d\mathbf{N}$ denotes the vector of a hypothetical AIM electron-population displacements from their equilibrium values $\mathbf{N}^0$. Since $d\mathbf{N} = -d\mathbf{N}_r$, the assumed equilibrium removes the first-order contribution to the associated change due to $\mathbf{z}$ in the energy, $E = E_M + E_r$, of the combined (closed) system (M∤r); moreover, taking into account the infinitely soft character of a macroscopic reservoir, the only contribution to the energy change in the quadratic approximation is:

$$E(\mathbf{z}) - E^0 = \tfrac{1}{2}\mathbf{z}\boldsymbol{\eta}\mathbf{z}^\dagger = dE(\mathbf{z}) \qquad (85)$$

where $E^0 = E(\mathbf{O})$ and $\boldsymbol{\eta} = (\partial^2 E/\partial \mathbf{z}\,\partial \mathbf{z})_0$ is the AIM hardness matrix of M. The equilibrium-restoring force is defined by the AIM electronegativities, relative to that of the reservoir:

$$\bar{\chi} \equiv -\frac{\partial E}{\partial z} = -\frac{\partial E_M}{\partial z} + \frac{\partial N}{\partial z}\frac{\partial E_r}{\partial N} = \chi - \chi_r \mathbf{1} \tag{86}$$

$$= (\chi^0 - z\eta) - \chi_r \mathbf{1} = -z\eta$$

where $\chi_r = -\mu_r$. The last equation allows one to express populational variables z in terms of the restoring force, $z = -\bar{\chi}\eta^{-1} = \bar{\mu}\sigma$, which identifies the softness matrix σ as the populational compliance matrix, $\sigma = \partial z/\partial\bar{\chi}$. Expressing the force field in terms of the force variables gives:

$$E(\bar{\chi}) - E^0 = \tfrac{1}{2}\bar{\chi}\sigma\bar{\chi}^\dagger \tag{87}$$

One can use this equation to define the k-th MEC,

$$\vec{\mathcal{N}}_k \equiv \sum_i^M (i)^k \bar{e}_k \equiv \bar{e}_M(\mathcal{N}^{(k)})^\dagger \tag{88}$$

where the k-th vector of population *interaction constants*, $\mathcal{N}^{(k)} = \{(i)^k\}$, is defined by the minimum energy criterion:

$$(i)^k \equiv (\partial z_i/\partial z_k)_{E=\min} \tag{89}$$

The explicit expression for these quantities immediately follows from Eq. (87), since the definition of Eq. (89) implies $\bar{\chi}_{i\neq k} = 0$. Thus:

$$(i)^k = (\partial z_i/\partial\bar{\chi}_k)/(\partial z_k/\partial\bar{\chi}_k) = \sigma_{i,k}/\sigma_{k,k} \tag{90}$$

The associated hardness (compliance constant) is defined by a similar minimum energy constraint:

$$\Gamma_k \equiv (\partial^2 E/\partial z_k^2)_{E=\min} = (\partial z_k/\partial\bar{\chi}_k)^{-1} = \sigma_{k,k}^{-1} \tag{91}$$

It represents the curvature of the section along $\vec{\mathcal{N}}_k$ of the energy function (Eq. (87)). This construction of the MEC is illustrated in Fig. 4, where one makes a unit populational shift along one AIM direction and searches for the minimum in the other AIM direction, identified by the point where the direction line is tangent to the corresponding ellipse contour of the force field. The figure shows that the MEC are not orthogonal, both being dominated by the softest PNM and internal polarization (P) components. The sum of the AIM displacements,

$$d\mathcal{N}^{(k)} \equiv \sum_i^M (i)^k = (\partial N/\partial N_k)_{E=\min} = \sigma_{k,k}^{-1}\sum_i^M \sigma_{i,k} \tag{92}$$

measures the global shift in N per unit displacement in N_k, when all remaining AIM are opened both internally (relative to the other AIM) and externally (relative to the reservoir).

Consider now the internal MEC. In this case we examine the closed M ($N = $ const.), with all AIM being free to exchange electrons among themselves. Let us again assume the initial global equilibrium, $\mu^0 = \mu^0\mathbf{1}$, and envisage the hypothetical polarizational displacements from N^0, $\bar{z} = (dN)_N$. The relevant compliants are now given by the internal softness matrix $S = (\partial N/\partial\mu)_N = -\beta$ (Eqs. 52–56), which defines the corresponding internal

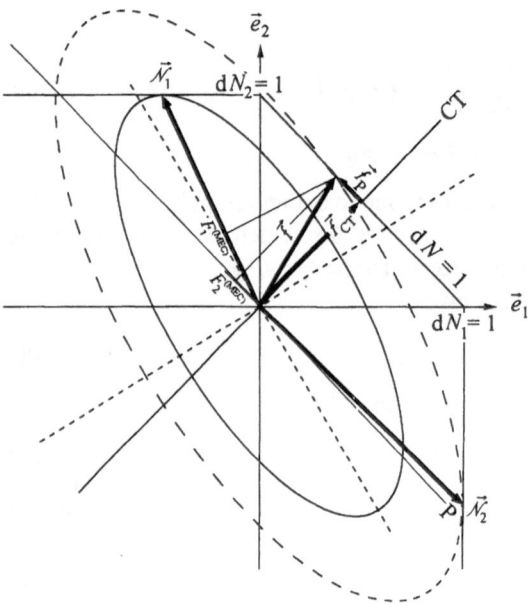

Fig. 4. The MECs $\{\vec{\mathcal{N}}_k, k = 1, 2\}$ in the case of the two-AIM populational degrees of freedom. Both MECs are seen to be dominated by the components from the soft PNM, $\bar{U}_2$ (see Fig. 1), or the corresponding projections onto the internal polarization axis, P, thus representing only minor changes in the overall number electrons (projections onto the CT direction)

MEC and their characteristics:

$$\vec{n}_k = \sum_i^M (\bar{i})^k \vec{e}_k \equiv \vec{e}_M (n^{(k)})^\dagger \tag{93}$$

$$(\bar{i})^k \equiv (\partial z_i / \partial z_k)_{N, E_M = \min} = (\partial \bar{z}_i / \partial \bar{z}_k)_{E_M = \min} = S_{i,k} / S_{k,k} \tag{94}$$

$$\gamma_k \equiv (\partial^2 E / \partial z_k^2)_{N, E_M = \min} = (\partial^2 E / \partial \bar{z}_k^2)_{E_M = \min} = S_{k,k}^{-1} \tag{95}$$

We would like to emphasize that, due to the closure constraint, there are only $(m - 1)$ linearly independent internal MEC. Thus, the m vectors defined by Eq. (93) in reality span the $(m - 1)$-dimensional space of internal MEC. In order to remove this linear dependence one could adopt the relative internal approach of Sect. 2.1.3. Namely, one then selects the electron population of one atom in the system as dependent upon populations of all remaining atoms, and discards the MEC associated with that atom. All remaining MEC can also be construc- ted directly from the corresponding internal relative softness matrix. Although the sets of independent internal MEC for alternative choices of the dependent atom will differ from one another, they must span the same $(m - 1)$-dimensional linear space of independent internal MEC. For example, in the two-AIM system of Fig. 4 there is only one independent internal MEC direction along the P-line.

It has been shown numerically that the above two sets of MEC, external and internal, are almost indistinguishable, exhibiting practically identical compliance constants of Eqs. (91) and (95). Therefore, in illustrative MEC plots for toluene (Fig. (5a)) we have displayed only the external MEC contours and listed their corresponding hardness parameters.

Clearly, it follows from the N = const. constraint,

$$d\bar{N}^{(k)} \equiv \sum_i^M (\bar{i})^k = (\partial N/\partial N_k)_{N, E = \min} = 0 \qquad (96)$$

that $\sum_{i \neq k}^M (\bar{i})^k = -(\bar{k})^k = -1$.

Analysis of the diagrams shown in Fig. (5a) reveals that the primary positive (unit) displacement $(k)^k = 1$, identified by the largest open circle in each contour, mainly occurs at the expense of the negative populational shifts on neighbouring atoms, which effectively screen the primary change, thus generating a rapidly decaying, oscillating response on remaining atoms. This explains why the $\bar{\mathcal{N}}_k$ and $\bar{n}_k$ MEC are almost identical. It is also worth noting that all contours exhibit the maximum sign alteration around the primary displaced atom. This indicates, that the softest PNM will always contribute strongly to all MEC.

Finally, let us define the REC. As we have mentioned earlier in this section the k-th REC, $\bar{\mathcal{R}}_k$, is defined by considering a partitioning of M into the displaced atom k, which accepts one electron from its reservoir, and the corresponding (closed) remainder, M'_k, a complementary subsystem

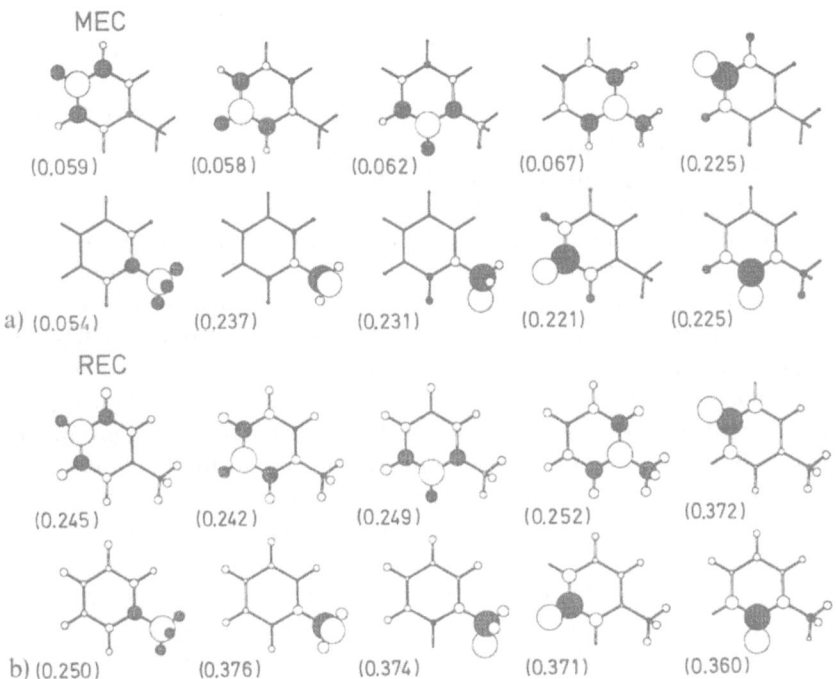

Fig. 5. Schematic diagrams of the external MEC (**Panel a**), $\bar{\mathcal{N}}_k$, and REC (**Panel b**), $\bar{\mathcal{R}}_k$, in toluene. The largest unshaded circle represents the atom k undergoing the test populational shift, $dN_k = 1$. The numbers in parentheses (a.u.) report the relaxed hardnesses Γ_k (**Panel a**) and $\eta_{k,k}^M$ (**Panel b**)

consisting of the mutually opened atoms different from k, $M'_k = (... |k - 1|k + 1|$
$...)$, $M^{(k)} = (k|M'_k)$:

$$\vec{\mathcal{R}}_k = \mathbf{\hat{e}}_k + \sum_{i \neq k}^{M} [i]^k \mathbf{\hat{e}}_i \equiv \mathbf{\hat{e}}_M (\mathcal{R}^{(k)})^\dagger \tag{97}$$

where $\mathcal{R}^{(k)} = (..., [k - 1]^k, 1, [k + 1]^k, ...)$ and

$$[i]^k \equiv (\partial z_i / \partial z_k)_{N'_k, E = min} \tag{98}$$

where $N'_k = \sum_{i \neq k}^{M} N_i \equiv \mathbf{1}'_k (N'_k)^\dagger$. Clearly, the electron flows of Eq. (98) are identical with those defined by the relaxational matrix of Eq. (39):

$$\mathcal{R}^{(k)} = \{[i]^k\} \equiv T^{(k|M'_k)} \equiv (\partial N'_k / \partial N_k)_{N'_k} \tag{99}$$

where $T^{(k|M'_k)} = \{T_{k,i}^{(k|M'_k)}, i \neq k\}$, so that $\vec{\mathcal{R}}_k$ is readily obtained from Eq. (42). Obviously, in the REC case: $\sum_{i \neq k}^{M} [i]^k = 0$. The hardness quantity associated with k-th REC is the AIM-relaxed hardness:

$$\eta_{k,k}^M = \eta_{k,k} + \sum_{i \neq k}^{M} \left(\frac{\partial N_i}{\partial N_k}\right)_{N'_k} \frac{\partial \mu_k}{\partial N_i} = \eta_{k,k} + \sum_{i \neq k}^{M} T_i^{(k|M'_k)} \eta_{i,k} \tag{100}$$

The illustrative REC for toluene are shown in Fig. (5b). A comparison between the corresponding modes of the two panels in the figure shows that they are very similar indeed, with the REC being seen to affect the atoms more distant from the location of the primary displacement; in its vicinity, where the amplitudes are the largest, both sets of contours are practically indistinguishable.

Analysis of the MEC hardnesses of Fig. (5a) shows that the test oxidations/reductions on carbon atoms give rise to comparable MEC hardnesses, the softest MEC in the system; the para (p)- and meta (m)- carbons in the ring are seen to be slightly softer in this respect than the ortho (o)- carbons. The softest population displacement is on the methyl carbon. The hard category of MEC involves the test oxidation/reduction on hydrogen atoms, with slightly harder changes being predicted for the methyl hydrogens. A similar picture emerges when one considers the relaxed (diagonal) atomic hardnesses reported in parentheses beneath the REC diagrams of Fig. (5b). Namely, the hardness differences between the (o-m-p)-carbons remains the same, but the methyl carbon appears almost as hard as the carbon that is chemically bonded to the methyl group. The softest hydrogen is now at the o-position, in contrast to the MEC hardness, which is predicted to be lowest at the m-location. Again, the methyl hydrogens are seen to be the hardest atoms in terms of the relaxed AIM hardness parameters.

Let us denote the MEC/REC directions by the representative vector $\vec{r}_M = (\vec{\mathcal{N}}_M, \vec{n}_M, \vec{\mathcal{R}}_M)$, where $\vec{\mathcal{N}}_M = (\vec{\mathcal{N}}_1, ..., \vec{\mathcal{N}}_m)$, etc., and define the corresponding representative transformation matrix $\mathbf{W} = (\mathcal{N}, n, \mathcal{R})$ between the basis of the orthonormal AIM vectors $\mathbf{\hat{e}}_M$ ($\mathbf{\hat{e}}_M^\dagger \mathbf{\hat{e}}_M = \mathbf{I}$) and the corresponding MEC/REC directions:

$$\vec{r}_M = \mathbf{\hat{e}}_M \mathbf{W} \tag{101}$$

The three transformation matrices $\mathbf{W}$ combine the MEC/REC components:

$$\mathscr{N} = (\mathscr{N}^{(1)} | \mathscr{N}^{(2)} | ... | \mathscr{N}^{(m)}) = \partial \vec{N}_M / \partial \mathring{e}_M \tag{102}$$

$$\boldsymbol{n} = (\boldsymbol{n}^{(1)} | \boldsymbol{n}^{(1)} | ... | \boldsymbol{n}^{(m)}) = \partial \vec{n}_M / \partial \mathring{e}_M \tag{103}$$

$$\mathscr{R} = (\mathscr{R}^{(1)} | \mathscr{R}^{(2)} | ... | \mathscr{R}^{(m)}) = \partial \vec{\mathscr{R}}_M / \partial \mathring{e}_M \tag{104}$$

It should be observed that, contrary to the orthonormal character of the AIM basis vectors, the MEC/REC vectors $\vec{r}_M$ are not orthogonal, giving rise to the metric

$$\mathscr{M} = \vec{r}_M^{\dagger} \vec{r}_M = \mathbf{W}^{\dagger} \mathring{e}_M^{\dagger} \mathring{e}_M \mathbf{W} = \mathbf{W}^{\dagger} \mathbf{W} \tag{105}$$

Clearly, the transformation $\mathbf{W}$ also defines the projections of the population displacement vector $\vec{z} = \sum_i^M dN_i \mathring{e}_i$ onto $\vec{r}_M = (\vec{r}_1, ..., \vec{r}_m)$:

$$z_k^{(r)} \equiv \vec{z} \vec{r}_k = \sum_i^M z_i W_{i,k} \quad \text{or} \quad z^{(r)} = z\mathbf{W} \tag{106}$$

Thus, the corresponding expression for the energy change in the quadratic approximation becomes:

$$\Delta E^{(1+2)} = \mu z^{\dagger} + \tfrac{1}{2} z \eta z^{\dagger}$$
$$= [\mu(\mathbf{W}^{-1})^{\dagger}](z^{(r)})^{\dagger} + \tfrac{1}{2} z^{(r)} [\mathbf{W}^{-1} \eta (\mathbf{W}^{-1})^{\dagger}](z^{(r)})^{\dagger} \tag{107}$$
$$\equiv \mu^{(r)}(z^{(r)})^{\dagger} + \tfrac{1}{2} z^{(r)} \eta^{(r)}(z^{(r)})$$

This equation defines the chemical potential $[\mu^{(r)}]$ and the hardness $[\eta^{(r)}]$ quantities for the MEC/REC reference frames. Similarly,

$$F^{(r)} \equiv \frac{\partial z^{(r)}}{\partial N} = \frac{\partial z}{\partial N} \frac{\partial z^{(r)}}{\partial z} = f\mathbf{W} \tag{108}$$

defines the relevant mode FF indices. Moreover, since $\mathring{e}^N = \mathring{e}_M \boldsymbol{1}^{\dagger} = \vec{r}_M(\mathbf{W}^{-1} \boldsymbol{1}^{\dagger})$, the complementary derivatives are:

$$(\boldsymbol{\varphi}^{(r)})^{\dagger} = \frac{\partial \mathring{e}^N}{\partial \vec{r}_M} = \frac{\partial N}{\partial z^{(r)}} = \mathbf{W}^{-1} \boldsymbol{1}^{\dagger} \tag{109}$$

These quantities vanish identically for the internal MEC, $\boldsymbol{\varphi}^{(n)} = \boldsymbol{0}$, and they are exactly equal to 1 in the REC case, $\boldsymbol{\varphi}^{(\mathscr{R})} = \boldsymbol{1}$, since then only the primary (unit) population displacement contributes to dN.

We observe further that the FF indices of Eq. (108) do not sum up to unity. However, following the PNM development one can again define the modified, CT-projected indices:

$$w_k^{(r)} = F_k^{(r)} \varphi_k^{(r)}, \quad k = 1, ..., m \tag{110}$$

which satisfy the desired "normalization":

$$F^{(r)}(\boldsymbol{\varphi}^{(r)})^{\dagger} = f\mathbf{W}\mathbf{W}^{-1} \boldsymbol{1}^{\dagger} = f\boldsymbol{1}^{\dagger} = 1 \tag{111}$$

The above observation calls for the corresponding CT-projected redefinition of the external MEC population variables:

$$\bar{z}^{(\mathcal{N})} = z\,\bar{\mathcal{N}}, \quad \bar{\mathcal{N}} = (\mathcal{N}^{(1)}\varphi_1^{(\mathcal{N})}|\mathcal{N}^{(2)}\varphi_2^{(\mathcal{N})}|\dots|\mathcal{N}^{(m)}\varphi_m^{(\mathcal{N})}) \tag{112}$$

and the associated derivatives:

$$\bar{\mu}^{(\mathcal{N})} = \mu(\bar{\mathcal{N}}^{-1})^{\dagger}, \quad \bar{\eta}^{(\mathcal{N})} = \bar{\mathcal{N}}^{-1}\eta(\bar{\mathcal{N}}^{-1})^{\dagger}, \quad \text{etc.} \tag{113}$$

It can be easily verified that $\bar{\mu}^{(\mathcal{N})} = \mu\mathbf{1}$ for global equilibrium state, when $\mu = \mu\mathbf{1}$. This is because

$$\bar{\varphi}_k^{(\mathcal{N})} \equiv \partial N/\partial\bar{z}_k^{(\mathcal{N})} = (\partial z_k^{(\mathcal{N})}/\partial\bar{z}_k^{(\mathcal{N})})\,(\partial N/\partial z_k^{(\mathcal{N})}) = \varphi_k^{(\mathcal{N})}/\varphi_k^{(\mathcal{N})} = 1 \tag{114}$$

In the internal MEC case, all populational derivatives $\mu^{(\pi)} = \mathbf{0}$ for the initial global equilibrium. Therefore, $\mu^{(\mathscr{R})} = \mathbf{0}$ also in the REC description for a given partitioning, $M^{(k)} = (k|M_k') = (k|i \neq k)$, if the initial electron distribution corresponds to the global equilibrium; then too, $\mu_k^{(\mathscr{R})} = \partial E/\partial z_k^{(\mathscr{R})} = \varphi_k^{(\mathscr{R})}\mu = \mu$. Thus, the representative REC chemical potential $\mu_k^{(\mathscr{R})}$ for the $M^{(k)}$ partitioning, representing the relaxed chemical potential of atom k, is also at the initial global chemical potential level.

The REC coefficient $[i]^k$ of Eq. (98) can be also expressed as the ratio of the corresponding softness quantities (see Eqs. (90) and (94)). Since the forces (internal electronegativities) associated with the relaxational degrees of freedom must vanish exactly by the minimum energy criterion, the following chain rule applies:

$$[i]^k = \left(\frac{\partial N_{i \neq k}}{\partial N_k}\right)_{N_k} = \left(\frac{\partial N_{i \neq k}}{\partial \mu_k}\right)_{N_k'} \bigg/ \left(\frac{\partial N_k}{\partial \mu_k}\right)_{N_k'} \tag{115}$$

where we have removed the $E = \min$ condition from the list of constraints in order to simplify the notation. The derivative in the numerator of the last expression is the off-diagonal softness for the $M^{(k)} = (k|M_k')$ partitioning of M:

$$S_{k,i}^{(k)} = \left(\frac{\partial N_k}{\partial \mu_k}\right)_{N_k'} \left(\frac{\partial N_{i \neq k}}{\partial N_k}\right)_{N_k'} = \frac{f_i^{k, M_k'}}{\eta_{k,k}^{M}} \tag{116}$$

defined by the ratio of the corresponding off-diagonal FF index and the relaxed hardness of the displaced atom k in M (Eq. (100)), the inverse of which also appears in the denominator of Eq. (115). Using the last two equations gives Eq. (99).

2.3 Mapping Relations

The ground-state relation between the geometric and electronic structure quantities provides a mapping between the nuclear-position and electronic-population degrees of freedom of molecular systems. An understanding of a subtle interplay between these parameters has been a major goal of theoretical chemistry [33–36]. It has recently been demonstrated [23, 37], that CSA provides a theoretical framework for constructing explicit *mapping relations* between the AIM (or collective) charge (electron population) coordinates and the corresponding nuclear (or collective) displacement modes. The crucial element of this development is the linear response (internal softness) matrix $\beta = -\mathbf{S}$, which links these two (electronic and geometrical) structural aspects of molecular systems in the AIM resolution.

Let $\mathbf{Q}$ denote the minimum set of bonds and angles that uniquely specify the system structure. The AIM-population displacements for constant N, $(d\mathbf{N})_N$, are linked with the corresponding geometrical displacements via the chain-rule transformation [23]:

$$(d\mathbf{N})_N = d\mathbf{Q}(\partial v/\partial\mathbf{Q})\beta \equiv d\mathbf{Q}\mathbf{G}\beta \equiv d\mathbf{Q}\mathbf{T}(\mathbf{Q} \to \mathbf{N})_N, \qquad (117)$$

where $\beta = (\partial\mathbf{N}/\partial v)_N$ and the *geometrical transformation* $\mathbf{G}$ represents the external potential response per unit change in the geometrical parameters. The latter is analytically available for the internuclear distances $\mathbf{Q} = \mathbf{R} \equiv (\mathbf{R}^b, \mathbf{R}^{nb})$, including the bonding (b) and nonbonding (nb) internuclear separations, respectively, in the "frozen" AIM-charges approximation, $q = q^0 = \mathbf{Z} - \mathbf{N}^0$, where $\mathbf{Z}$ groups the nuclear atomic numbers and $\mathbf{N}^0$ are the equilibrium AIM populations at the starting geometric structure $\mathbf{Q}^0$.

The charges from EEM calculations [10] were shown to provide an adequate basis for the model force field which determines a realistic bond-stretch Hessian matrix, $\mathscr{H}^b = \partial^2 E/\partial\mathbf{R}^b\partial\mathbf{R}^b|_{\mathbf{Q}^0}$, calculated for the fixed bond angles at the starting geometry [23]. For each atomic pair, this force field uses the effective point-charge electrostatic interactions as the potential component contribution of this pair to the total Born-Oppenheimer energy, and supplements it with a matching electronic kinetic energy function, obtained from the virial theorem, in the spirit of the familiar simple-bond-charge model [38]. Thus, in this approximation one attributes the Fues potential-like contribution of each atomic pair (bonding or nonbonding) to the overall potential energy surface for nuclear motions. The full geometric Hessian, $\mathscr{H} = \partial^2 E/\partial\mathbf{Q}\partial\mathbf{Q}|_{\mathbf{Q}^0}$, or $\mathscr{H}^b$ can also be generated from independent SCF MO/CI or DFT calculations.

Clearly, the $d\mathbf{N}$ and $d\mathbf{Q}$ vectors have different dimensions. Suppose that one is interested in relating the $(m - 1)$ populational degrees of freedom for constant N, $(d\mathbf{N})_N$ to the equidimensional subset, $\mathbf{Q}^s$ of $\mathbf{Q}$, e.g., $\mathbf{Q}^b \equiv \mathbf{R}^b$, in noncyclic systems. In order to fix bond angles one requires an additional *constraint transformation*, $\tau^s = \partial\mathbf{Q}/\partial\mathbf{Q}^s$, which is available from geometrical considerations [23]. For example, the $d\mathbf{R}^b \to (d\mathbf{N})_N$ mapping is given by the product (chain rule) transformation:

$$(d\mathbf{N})_N = d\mathbf{R}^b\tau^b\mathbf{T}(\mathbf{Q} \to \mathbf{N})_N \equiv d\mathbf{R}^b\mathbf{T}(\mathbf{R}^b \to \mathbf{N})_N. \qquad (118)$$

These *local mapping transformations*, relating AIM electron populations (charges) to bond lengths, can be easily generalized into relations involving collective electron-population- and/or nuclear-position-displacements, e.g., PNM and nuclear normal modes. For example, the bond-stretching normal vibrations, $\mathscr{Q}^b$, defined by the $\mathscr{H}^b$ principal directions, $\mathbf{O} = \partial\mathscr{Q}^b/\partial\mathbf{R}^b$,

$$\mathbf{O}^{\dagger}\mathscr{H}^{b}\mathbf{O} = \mathbf{k}^{b} = \{k_{m}^{b}\,\delta_{mn}\}, \quad \mathbf{O}^{\dagger}\mathbf{O} = \mathbf{I}, \tag{119}$$

are related to the AIM charges through the $d\mathscr{Q}^{b} \rightarrow (d\mathbf{N})_{N}$ mapping:

$$(d\mathbf{N})_{N} = d\mathscr{Q}^{b}\mathbf{O}^{\dagger}\mathbf{T}(\mathbf{R}^{b} \rightarrow \mathbf{N})_{N} \equiv d\mathscr{Q}^{b}\mathbf{T}(\mathscr{Q}^{b} \rightarrow \mathbf{N})_{N} \tag{120}$$

Consider now the $(m - 1)$ collective internal PNM (internal normal modes, INM), represented by all eigenvectors of $\boldsymbol{\beta}$ except the zero-eigenvalue $(d\mathbf{N}$, external CT) mode $\alpha = m$, for which all components are identical:

$$\mathbf{c}^{\dagger}\boldsymbol{\beta}\mathbf{c} = \mathbf{s} = \{s_{\alpha}\delta_{\alpha\gamma}\}, \quad \mathbf{c}^{\dagger}\mathbf{c} = \mathbf{I} \tag{121}$$

The transformation $\mathbf{c}$ can be partitioned into the matrix $\mathbf{c}'$, grouping the first $(m - 1)$ columns and representing independent internal charge redistribution modes, and the last column, representing the pure CT displacement: $\mathbf{c} = (\mathbf{c}', \mathbf{c}_{m})$. The $(m - 1)$ collective populational variables,

$$d\mathbf{p}' = (d\mathbf{N})_{N}\mathbf{c}' \tag{122}$$

can now be related to geometrical degrees of freedom. For example, multiplying Eq. (118) by $\mathbf{c}'$ immediately gives the $\mathbf{T}(\mathbf{R}^{b} \rightarrow \mathbf{p}')_{N}$ mapping:

$$d\mathbf{p}' = d\mathbf{R}^{b}\mathbf{T}(\mathbf{R}^{b} \rightarrow \mathbf{N})_{N}\mathbf{c}' \equiv d\mathbf{R}^{b}\mathbf{T}(\mathbf{R}^{b} \rightarrow \mathbf{p}')_{N}. \tag{123}$$

Finally, expressing $d\mathbf{R}^{b}$ in tems of $d\mathscr{Q}^{b}$, $d\mathbf{R}^{b} = d\mathscr{Q}^{b}\,\mathbf{O}^{\dagger}$, gives the $d\mathscr{Q}^{b} \rightarrow d\mathbf{p}'$ mapping transformation:

$$d\mathbf{p}' = d\mathscr{Q}^{b}\mathbf{O}^{\dagger}\mathbf{T}(\mathbf{R}^{b} \rightarrow \mathbf{p}')_{N} \equiv d\mathscr{Q}^{b}\mathbf{T}(\mathscr{Q}^{b} \rightarrow \mathbf{p}')_{N}. \tag{124}$$

Obviously, one can derive similar mapping relations involving the populational and/or nuclear MEC or REC. However, when generating the relevant transformations one has to recognize their non-orthogonal character.

In Fig. (6) we have collected illustrative mapping "translators" between the geometrical (bond-stretch) and populational degrees of freedom in $CH_{3}OH$ [23]. These were obtained from the MNDO bond-stretch Hessian matrix and the relevant charge-sensitivity data from the CSA/EEM calculations in the AIM resolution.

The first panel (a1) shows the bond length responses per unit populational displacements (chemical reduction, $dN_{i} = 1$) of the constituent AIM. It can be seen in the diagram that such test displacements on methyl hydrogens (reversing original bond polarization) are predicted to result in elongation of both the CO (secondary response) and the CH (primary response) bonds. A similar direct geometrical relaxation, stretching the CO bond, is predicted when the electron population on C is increased. The reverse, bond shortening, geometric response is predicted when the hydroxyl atoms are reduced: for both cases $(i = 1, 3)$. The major relaxations involve a dominant increase in R_{CO}^{b}. There is also a secondary structural change of $dR_{CH}^{b} < 0$ accompanying a reduction of the OH group. The assumed chemical reduction on the oxygen atom implies an increase in the initial polarity of the CO bond, so that the a1 diagram associates the CO-bond

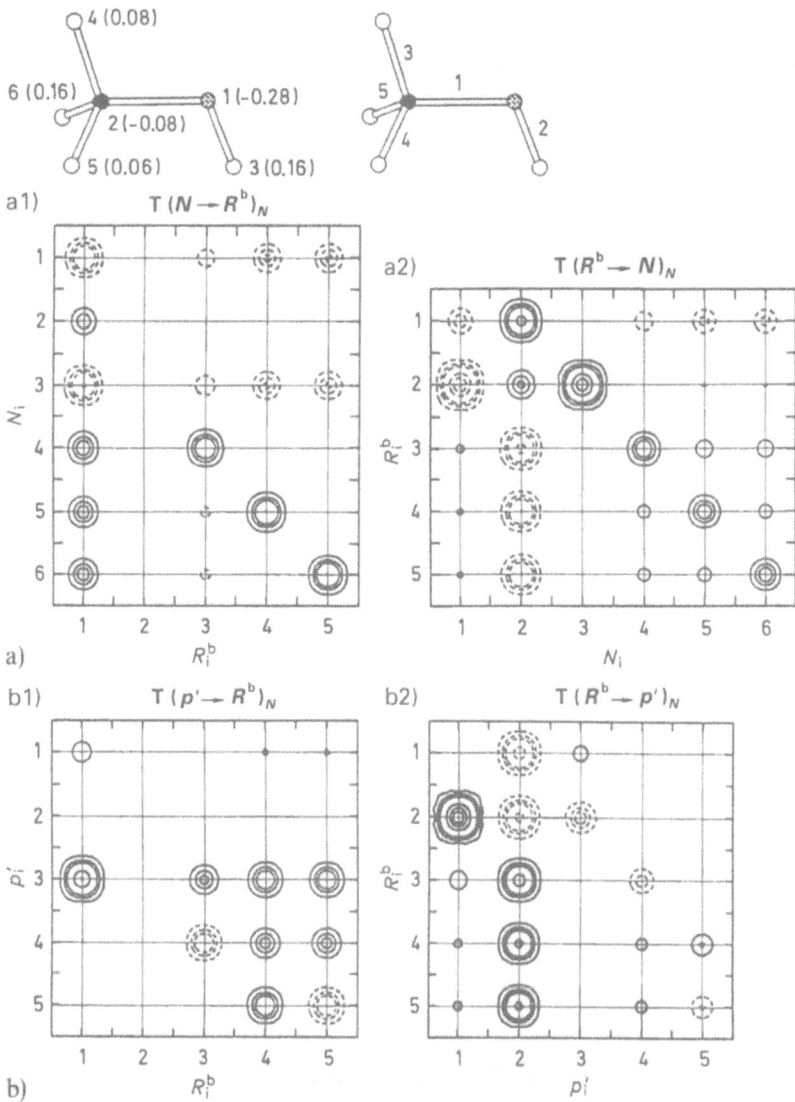

Fig. 6. (Continued).

shortening with increased bond polarity. The same physical implications follow from the $dN_C > 0$ displacement, which decreases the initial CO polarity and, as such, is related to the CO-bond-lengthening response. One could interpret the CH-bond-length responses to populational (charge) shifts on methyl hydrogens in the same spirit: test reductions on the hydrogen were shown to decrease bond polarity, and therefore, are predicted to stretch the CH bonds. Finally, one concludes from the a1 map that the OH bond remains relatively resistant to all

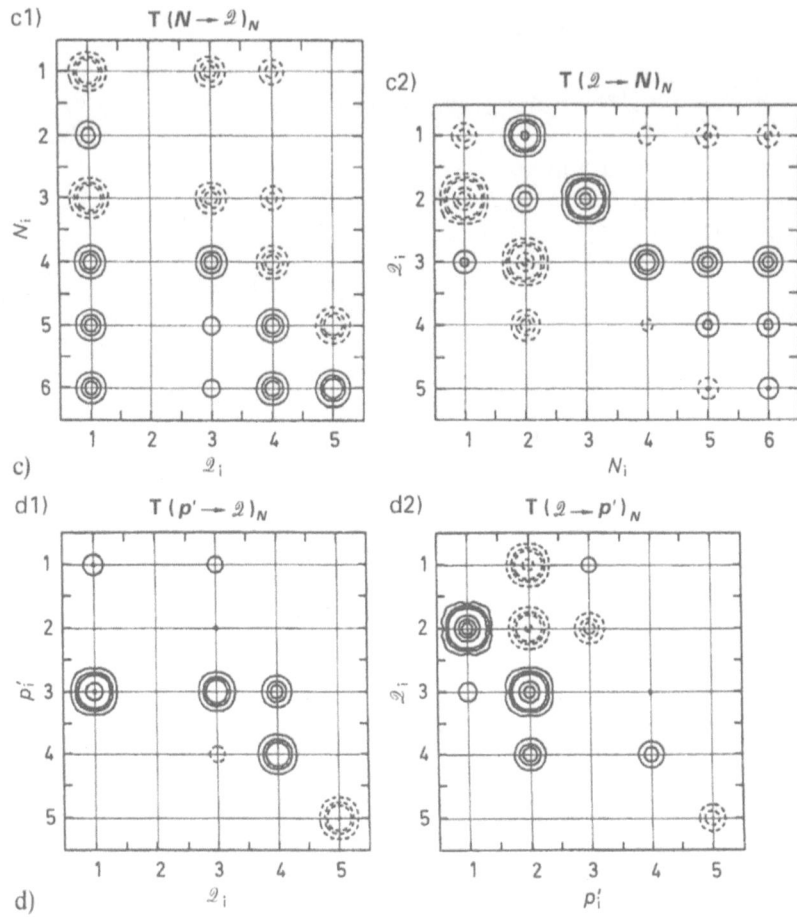

Fig. 6. (Continued).

AIM charge displacements. Clearly, all of the above responses are reversed when the test chemical oxidations ($dN_i < 0$) are considered.

Let us now examine the inverse mapping transformation (Panel a2), which has a direct connection to the Gutmann *bond-length variation rules* [33]. This map exhibits the dominant and intuitivity expected *direct charge response effect*, i.e., decreasing polarity of the elongated bond. For example, a test stretch of the CO bond mainly increases the electron population on C and decreases it on O, thus resulting in a lower bond polarity. Similarly, $dR_{OH}^b > 0$ generates $dN_O < 0$ and $dN_{H(O)} > 0$, thus diminishing the initial bond polarization. A similar, though weaker, effect is detected in all CH bonds, in which the electron population of the carbon atom decreases and that of the corresponding hydrogen increases when the bond is elongated. All these primary charge-relaxation

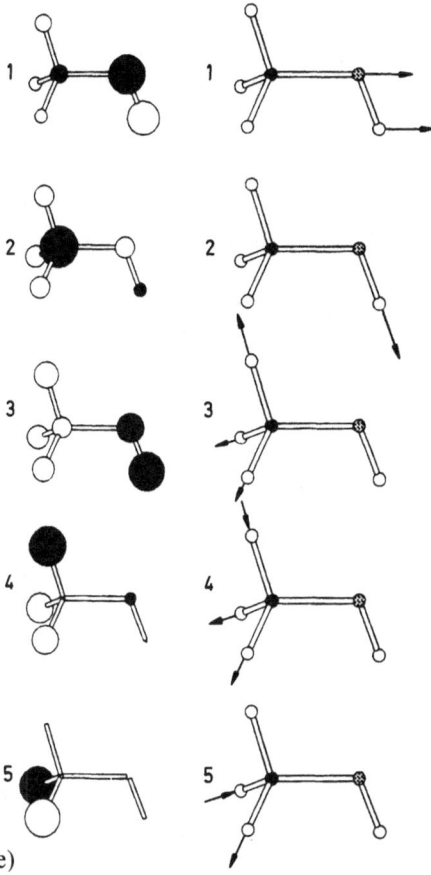

Fig. 6. Illustrative mapping relations [from Ref. 23] between the populational and bond-stretching degrees of freedom in methanol: local-local (**Panels a**), collective-local (**Panels b**), local-collective (**Panels c**), and collective-collective (**Panels d**). The corresponding local and collective structure variables as well as the initial EEM charges are defined in additional diagrams of **Panels a** and **e**. The matrix "translators" $T(x \to y)$ are interpreted as a "surface" and graphically represented as the corresponding contour maps, with the row (independent) and column (dependent) variables x and y fixed along the horizontal and vertical lines, respectively. In **Panel e** the same convention as that previously used to illustrate collective charge displacements (first column) has been adopted, with the circle areas being proportional to $|c_{i\alpha}|$ (Eq. (121)); the lengths of arrows in the second column of **Panel e** is proportional to the dominating $|O_{rn}|$ components (Eq. (119)) in the n-th vibrational normal mode. There are two sets of contour values, used for panels in each column: in the first column (al-dl) the starting value of $|T_{kl}| = 2$ with equal spacing between 3 consecutive contours, $\Delta = 2$; above $|T_{kl}| = 6$ $\Delta = 10$ has been used; the contour values in the second column are divided by 100, relative to the first column. Negative values are represented by broken lines

effects are reversed when the $dR_i^b < 0$ primary structural displacements are examined.

There are several weak indirect bond-polarization responses in a 2 map of Fig. (6). Namely, the CO stretch is seen to diminish the original polarity of the

OH bond, since the net charge on O decreases and that on the hydroxyl hydrogen remains approximately constant. Similarly, the OH stretch is predicted to result in diminished initial CO bond polarity, since it implies the secondary responses: $\delta N_C > 0$ and $\delta N_O < 0$.

In non-cyclic systems each bond, i, divides the molecule into two complementary sybsystems. Let b_i^D and a_i^A denote the donor (basic) and acceptor (acidic) atoms of the heteronuclear, coordination bond $b_i^D \rightarrow a_i^A$; let D_i and A_i stand for the donor (consisting b_i^D) and acceptor (including a_i^A) subsystems, respectively:

$$
\begin{array}{cccc}
\overbrace{}^{D_i} & & \overbrace{}^{A_i} & \\
\cdots a_i^D \text{-----} b_i^D \text{-------} a_i^A \text{-------} b_i^A \cdots \\
\delta - & \delta + & \delta - & \delta +
\end{array}
$$

(125)

Here a_i^D and b_i^A represent the remaining acidic and basic atoms, respectively, in D and A bonded to the two coordinating atoms of the perturbed bond, generally exhibiting fractional negative and positive charges, respectively, as a result of the intra-D_i and intra-A_i bonds formation. The intuitively expected electron distribution response to a lengthening of bond i can be summarized by the following Gutmann-like diagram:

$$
\cdots a_i^D \leftarrow \text{---} b_i^D \leftarrow \text{------} a_i^A \text{----} \rightarrow b_i^A \cdots
$$
$$
D_i \qquad dR_i^b > 0 \qquad A_i
$$

(126)

where the arrows connecting the bonded nuclei point toward the induced electron shift (bent arrows) or an increase in a bond length. This diagram takes into account a decreased CT component of the bond being stretched (the directly coupled response) and the polarizations inside D_i and A_i (indirectly coupled, induced responses), which moderate the effects of the primary charge shift in the Le Châtelier-Braun sense.

Paraphrasing the first bond-length-variation rule, which summarizes changes in the molecule D and A when the intermolecular $D \rightarrow A$ distance is increased, suggests the following general behaviour: the larger the distance R_i^b, the smaller the polarization of the stretched bond and the greater the coordination (shorter bond) of bonds involving b_i^D and a_i^A. As a result of this charge redistribution, the a_i^D atom becomes less acidic and the b_i^A atom becomes less basic. However, since the direct-response charge redistributions are expected to be larger than the polarizational ones, the outflow (inflow) of electrons from a_i^A (to b_i^D) when R_i^b is increased should be larger than the moderating (induced) inflow (outflow) due to remaining atoms. This also suggests a decrease in the initial bond polarity of the other bonds involving constituent atoms of the primarily stretched bond.

Some of these qualitative predictions are indeed observed in Panel (a2) of Fig. (6). For example, as we have already observed above, the $dR_2^b > 0$ perturbation of the OH bond does indeed result in decreased CO bond polarity; a similar effect is observed for the OH bond when R_{CO}^b is increased.

We now turn to the mixed translators of Panels b. The first of them (b1) maps the collective charge modes (INM, Panel e) into the bond length variations that they imply. The modes $\alpha = 1, 2$ are seen to have relatively weak effect upon all bonds. The $\alpha = 3$ mode, describing the $[OH] \rightarrow [CH_3]$ internal CT, has the strongest geometry-relaxation response, in which all bonds but OH are predicted to increase their lengths as a result of the forced charge displacement. Finally, the remaining internal CT modes between methyl hydrogens, $\alpha = 4, 5$, are seen to give rise to CH-bond differentiation, as intuitively expected.

The inverse mapping, from the forced bond stretches to the INM displacements (Panel b2), shows that the CO stretch in methanol translates almost exclusively into a displacement of the second INM in the negative direction, if one neglects a small positive component from $\alpha = 3$. The $\alpha = 2$ INM is seen to participate strongly in charge responses to all of the bond elongations. This is not surprising, since this softest mode (exhibiting the largest number of nodes) represents the very (short-range) polarization which is responsible for the minimum-energy charge reorganization following a given bond-length increase. One also concludes from this diagram that the localized OH stretch requires a combination of displacements along the $\alpha = 1 - 3$ INM to ascribe the charge response to such a geometrical perturbation. This is consistent with the local mapping relations. Namely, in order to account for direct and indirect charge responses to an increase in R_{OH}^b one needs the OH polarization (mode $\alpha = 1$) and CO (CH) polarization (modes $\alpha = 2, 3$) components. This is also the reason why the softest mode $\alpha = 2$, that includes alternating polarizations between neighbouring atoms, is present so strongly in all charge responses to localized bond stretches.

Now, let us examine the mixed translators of Panel c, involving the bond-stretch normal modes (Panel e), and the AIM electron populations (shown above Panel a). We observe first, that the localized nature of the first two nuclear modes (essentially localized CO and OH stretches), implies that the two first columns of the c1 translator should be identical with those in Panel a1, as indeed seen in the Figure. It should be emphasized at this point that such local (test) reduction of a given atom i generates a charge response along the i-th internal populational MEC, which triggers the structural change represented by diagrams a1 and c1. Thus, the structural responses reflected by these mapping relations do in fact represent the geometry relaxation effects associated with the populational (internal) MEC in question. A comparison between diagrams a1 and c1 shows that the local charge perturbations in general translate into a combination of many vibrational normal modes.

Turning now to transformation c2, we again observe that the first two rows are almost identical with those in map a2, again due to the localized character of $\mathcal{Q}_1^b$ and $\mathcal{Q}_2^b$. The symmetric stretch of the CH bonds $(\mathcal{Q}_3^b)$ is again seen to result in their diminished polarity; a slight increase in the polarity of the CO bond is also observed in response to a shift along this collective nuclear displacement. Again, one detects in the asymmetric stretches $\mathcal{Q}_4^b$ and $\mathcal{Q}_5^b$ that always the stretched (shortened) CH bonds are mapped into decreased (increased) bond polarity changes.

The most compact mapping is obtained in Panel d1, in which displacements along populational INM are translated into normal bond-stretches. It follows from this diagram that the softest INM does not generate any appreciable movement of nuclei along five nuclear coordinates. This prediction is consistent with our earlier observations relating the polarity of a bond to its length. Namely, since this INM changes the polarity of all bonds it cannot translate into a nuclear mode representing only localized CH_3 bond stretches. The $\alpha = 1$ INM increases the CH(4) and OH bond lengths slightly, the polarities of which decrease as a result of a positive INM displacement. The third INM, which shifts electrons from OH to CH_3, is seen to be mapped into a combination of positive displacements along $\mathcal{Q}_1^b$, $\mathcal{Q}_3^b$, and $\mathcal{Q}_4^b$. Again the $\mathcal{Q}_1^b$ (CO stretch) component dominates in accordance with the most important decrease in the CO polarity along this INM. The remaining components can also be explained in terms of this approximate bond polarity – bond length relationship. Namely, the $\alpha = 3$ INM is now seen to generate decreased polarity of the non-bonded OH(4,5) interactions. Thus, this electron redistribution should also translate into an appreciable increase in the length of bonds 4 and 5, and a smaller increase in R_3^b. This is indeed the net displacement of the methyl hydrogens represented by the ($\mathcal{Q}_3^b$, $\mathcal{Q}_4^b$)-components in the $\alpha = 3$ row of Panel d1. Finally, the $\alpha = 4$ and 5 INM translate into $\mathcal{Q}_4^b$ and $\mathcal{Q}_5^b$ almost on the one-to-one basis. These mappings are partly due to the symmetry constraints and they conform fully to the general bond-polarity-bond length relation.

Finally, let us interpret the physical implications that follow from the inverse collective-collective transformation of Panel d2. It relates the independent (unit) displacements along the nuclear normal modes into the respective charge responses in the INM representation. As expected, the $\mathcal{Q}_1^b$ (CO stretch) mode is mapped into the minimum energy response represented by the dominant (negative) shift along $\alpha = 2$ INM, as is required in order to decrease the CO bond polarity. A more complicated charge shift accompanies unit displacement along $\mathcal{Q}_2^b$ (OH stretch). Clearly, the dominant INM in this nuclear motion, $\alpha = 1$, accounts for the decreased OH polarity. The remaining, negative shifts along $\alpha = 2$ and 3 INM are required to represent the short- and long-range polarization, describing charge responses on remaining atoms, which counteract the charge-reorganization effects, due to $\alpha = 1$ component, that are conjugated directly to the $d\mathcal{Q}_1^b > 0$ displacement. This Le Châtelier-Braun type interpretation can also rationalize the small component in the next $\mathcal{Q}_3^b$ (symmetric CH_3 stretch) row of Panel d2. In this case, the directly coupled normal-charge response, $\alpha = 2$, which dominates the charge relaxation, describes mainly a collective charge shift that diminishes the initial bond polarity in the CH_3 group; a small positive component from $\alpha = 1$ INM accounts for the induced polarizational shift inside the OH bond. The intuitively expected INM charge reorganization spectrum is obtained for the geometry change along $\mathcal{Q}_4^b$, since a combination of positive shifts along $\alpha = 2$ and $\alpha = 4$ generates the required CH bond polarity changes for CH(5,6) (i.e., stretching the bonds). Finally, due to symmetry, $\mathcal{Q}_5^b$ is seen to be translated into a single conjugated shift along $\alpha = 5$ INM.

All chemical reactions involve both the nuclear displacements and the concomitant electron redistribution. By the Hohenberg-Kohn theorems [1] the ground state electron density is in one-to-one correspondence to the underlying external potential due to the nuclei of the system. Depending on what is considered to be the perturbation and what the response to it, when the geometry of the system undergoes a given change, either of the following two perspectives can be adopted [36] : (1) *electron-cloud following*, when the electron density shift is viewed as the ground-state *response* to the primary nuclear displacement (*perturbation*), in the spirit of the Born-Oppenheimer approximation; (2) *electron-cloud preceding*, in which a hypothetical electron density displacement (*perturbation*) is regarded as preceding (accelerating) changes in nuclear positions (*response*). It is possible, under simplifying assumptions, to rationalize some electron processes in terms of the MO transition densities, although many transition densities are expected to contribute even to the simplest responses of the system [34, 35]. Symmetry rules for chemical reactions have also been formulated [34, 35] in terms of symmetry restrictions on the excited state contributions to the ground-state wave-function, due to perturbation by a given normal mode displacement of the nuclei. The mapping relations of the CSA in the AIM resolution provide a more explicit and transparent framework for discussing the general, ground-state relations between the *Electron Population Space* (EPS) and the *Nuclear Position Space* (NPS). There is an urgent need for means to "*translate*" directly the reaction coordinate representations in the two spaces into one another. As we have demonstrated above, the CSA development generates these very "translators" between attributes of alternative reference frames in the two spaces.

The explicit mapping relations have profound implications for catalysis and for the theory of chemical reactivity in general, since a given, desired geometrical (bond-stretch) NPS displacement of a system from its initial structure can now be translated into the associated charge EPS shifts, and *vice versa*. One can also transform selected reactive EPS channels, e.g., the softest PNM or INM of reactants, into their NPS conjugates. The numerical data reported for methanol also provide additional validation of the physical significance of the PNM reference frame, which is practically indistinguishable from the INM coordinate system. The nuclear normal modes are known to have physical meaning at low energies (amplitudes) (*cf.* "*natural vibrations*"); they are now almost uniquely related to their EPS conjugates (*cf.* "*natural polarizations*"): the geometrical/electronic structure relaxations of the system are seen to take place along a very few normal modes, in both the NPS and EPS representations. The one-to-one character of some normal mapping relations is also important for interpretative (predictive) purposes. The successful applications of the PNM reactivity criteria can be given a more conventional NPS rationalization when the collective-collective translators are dominated by one-to-one relations: electrons in such molecular systems do indeed tend to be redistributed along a single INM/PNM in response to a geometrical distortion along a single conjugated normal coordinate.

Both the electron-cloud preceding (that accelerates nuclear motion) and electron-cloud incomplete following (resisting nuclear motion) mechanisms of Nakatsuji [36] can now be viewed as particular manifestations of the present mapping relations.

The possibility of mapping local (or collective) displacements in one space onto the local (or collective) shifts in the other space can be used to quickly assess the feasibility of various populational displacements in a molecule, by testing whether they translate into one of the normal vibrations. One can also achieve a direct connection between the charge sensitivities and spectroscopic data, which might eventually help one to establish additional ways of verifying the reactivity trends predicted within CSA. These mapping relations also provide a sound phenomenological basis for a systematic manipulation of surface structures, in order to facilitate the desired charge redistribution in the adsorbate. This opens new possibilities for affecting reactivities and selectivities along a given reaction channel.

One could use the NPS-EPS translators to identify natural vibration modes associated with a given normal or local polarization channels. This is potentially important for catalysis, in which polarizational promotion of an adsorbate molecule by an active site of the surface can eventually be translated into bond-stretch vibrations, perhaps those leading to bond cleavages. One could also attempt a search for the optimum matching of two reactants with the aim of achieving the optimum charge polarizations, ultimately leading to the desired bond-forming – bond-breaking mechanism.

As emphasized by Gutmann [33], the structure of a molecular system is not rigid, but rather is adaptable to environment. The CSA/EEM mapping relation development allows one to view this electronic-geometrical structure relationship from many chemically interesting perspectives. Changes in the charge density pattern are reflected by changes in the structural pattern, which in turn may lead to a new reactivity pattern in the system. By applying an appropriate molecular environment, e.g., a catalyst, one can induce the desired electronic structural conditions for "activating" bonds of interest by changing their polarity.

3 Molecular fragment development

3.1 Determining Charge Sensitivities from Canonical AIM Data

One can determine the fragment responses corresponding to the constrained equilibrium in $M^G = (X|Y|...|Z)$ in an analogous way to that used to determine the responses and global properties of the system as a whole (Eqs. (57) and (58)). In this case the system is divided into mutually closed fragments, each exhibiting its own equilibrium chemical potential: $\mu_X = \mu_X^M \mathbf{1}_X, ..., \mu_Z = \mu_Z^M \mathbf{1}_Z$. The electronegativity-equalization equations, which describe the intra-fragment rearrangement leading to the assumed equilibrium, together with the obvious relations, $\{\sum_i^X dN_i = dN_X\}$, may be written in the following matrix form [20]:

$$
\begin{bmatrix}
0 & 0 & \cdots & 0 & \mathbf{1}_X & \mathbf{0}_Y & \cdots & \mathbf{0}_Z \\
0 & 0 & \cdots & 0 & \mathbf{0}_X & \mathbf{1}_Y & \cdots & \mathbf{0}_Z \\
\cdots & \cdots & \cdots & \cdots & \cdots & \cdots & \cdots & \cdots \\
0 & 0 & \cdots & 0 & \mathbf{0}_X & \mathbf{0}_Y & \cdots & \mathbf{1}_Z \\
\mathbf{1}_X^\dagger & \mathbf{0}_X^\dagger & \cdots & \mathbf{0}_X^\dagger & \eta^{X,X} & \eta^{X,Y} & \cdots & \eta^{X,Z} \\
\mathbf{0}_Y^\dagger & \mathbf{1}_Y^\dagger & \cdots & \mathbf{0}_Y^\dagger & \eta^{Y,X} & \eta^{Y,Y} & \cdots & \eta^{Y,Z} \\
\cdots & \cdots & \cdots & \cdots & \cdots & \cdots & \cdots & \cdots \\
\mathbf{0}_Z^\dagger & \mathbf{0}_Z^\dagger & \cdots & \mathbf{1}_Z^\dagger & \eta^{Z,X} & \eta^{Z,Y} & \cdots & \eta^{Z,Z}
\end{bmatrix}
\begin{bmatrix}
-\mu_X^M \\
-\mu_Y^M \\
\cdots \\
-\mu_Z^M \\
\delta N_X^\dagger \\
\delta N_Y^\dagger \\
\cdots \\
\delta N_Z^\dagger
\end{bmatrix}
$$

$$
= \begin{bmatrix}
dN_X^M \\
dN_Y^M \\
\cdots \\
dN_Z^M \\
-(\mu_X^+)^\dagger \\
-(\mu_Y^+)^\dagger \\
\cdots \\
-(\mu_Z^+)^\dagger
\end{bmatrix}, \tag{127}
$$

or in short: $\tilde{\eta}_G \mathcal{R}_G^\dagger = \mathcal{P}_G^\dagger$

where $\mathcal{R}_G$ and $\mathcal{P}_G$ stand for the generalized response and perturbation vectors in M^G, respectively. As before in the case of global equilibrium, the vectors, $\mu_A^+ = \mu_A^0 \mathbf{1}_A + dv_A (A = X, Y, ..., Z)$ group the initially displaced chemical potentials of constituent atoms that define the intra-fragment forces, $d\mu_A^+ = \mu_A^+ - \mu_A^M \mathbf{1}_A$ $(A = X, Y, ..., Z)$, and thus determine the changes in the electron populations of the atoms, $\delta N_A = N_A - N_A^0$, etc., as allowed by the partitioning M^G. Again, the solution of this equation can be found by the inversion of the generalized hardness matrix $\tilde{\eta}_G$:

$$
\begin{pmatrix}
-\mu_X^M \\
-\mu_Y^M \\
\cdots \\
-\mu_Z^M \\
\delta N_X^\dagger \\
\delta N_Y^\dagger \\
\cdots \\
\delta N_Z^\dagger
\end{pmatrix}
=
\begin{pmatrix}
-\eta_{X,X}^M & -\eta_{X,Y}^M & \cdots & \eta_{X,Z}^M & f^{X,X} & f^{X,Y} & \cdots & f^{X,Z} \\
-\eta_{Y,X}^M & -\eta_{Y,Y}^M & \cdots & \eta_{Y,Z}^M & f^{Y,X} & f^{Y,Y} & \cdots & f^{Y,Z} \\
\cdots & \cdots & \cdots & \cdots & \cdots & \cdots & \cdots & \cdots \\
\eta_{Z,X}^M & \eta_{Z,Y}^M & \cdots & \eta_{Z,Z}^M & f^{Z,X} & f^{Z,Y} & \cdots & f^{Z,Z} \\
(f^{X,X})^\dagger & (f^{Y,X})^\dagger & \cdots & (f^{X,Z})^\dagger & -\beta^{X,X} & -\beta^{X,Y} & \cdots & -\beta^{X,Z} \\
(f^{X,Y})^\dagger & (f^{Y,Y})^\dagger & \cdots & (f^{Z,Y})^\dagger & -\beta^{Y,X} & -\beta^{Y,Y} & \cdots & -\beta^{Y,Z} \\
\cdots & \cdots & \cdots & \cdots & \cdots & \cdots & \cdots & \cdots \\
(f^{X,Z})^\dagger & (f^{Y,Z})^\dagger & \cdots & (f^{Z,Z})^\dagger & -\beta^{Z,X} & -\beta^{Z,Y} & \cdots & -\beta^{Z,Z}
\end{pmatrix}
\begin{pmatrix}
dN_X^M \\
dN_Y^M \\
\cdots \\
dN_Z^M \\
-(\mu_X^+)^\dagger \\
-(\mu_Y^+)^\dagger \\
\cdots \\
-(\mu_Z^+)^\dagger
\end{pmatrix},
$$

$$
\mathscr{R}_G^\dagger = \tilde{\sigma}\,\mathscr{P}_G^\dagger, \tag{128}
$$

where $\tilde{\sigma}_G = (\tilde{\eta}_G)^{-1}$ is the molecular fragment generalized softness matrix. It can be partitioned into two main diagonal blocks, representing the condensed hardness matrix elements and those of the linear response function, respectively, with the off-diagonal blocks giving the relaxed FF indices (diagonal and off-diagonal). For each fragment one obtains the familiar equation $d\mu_A^M = \sum_B^M dN_B^M \eta_{A,B}^M + \sum_B^M dv_B(f^{A,B})^\dagger$, which generates the shift in the chemical potential of fragment A, and the second equation $\delta N_A = \sum_B^M dN_B^M f^{B,A} + \sum_B^M dv_B^M \beta^{B,A}$ (A, B = X, Y, ..., Z), for the relevant intra-fragment charge reorganization. Here, the matrix blocks of the FF and linear response matrices are defined as follows:

$$
f^{X,Y} \equiv \partial N_Y/\partial N_X = \partial\mu_X/\partial v_Y, \quad \beta^{X,Y} \equiv \partial N_Y/\partial v_X.
$$

Using the chain-rule transformation, one can generalize Eq. (43) for this general case of constrained equilibrium:

$$
f^{A,B} = \frac{\partial N_B}{\partial N_A} = \sum_C^M \frac{\partial N_C}{\partial N_A}\frac{\partial N_B}{\partial N_C} \equiv \sum_C^M f^{A,C} T^{(C|B)} \tag{129}
$$

where $T^{(C|B)}$ is a block of the relaxational matrix of Eqs. (39) and (42), generalized for partitioning a molecule into molecular fragments. Knowledge of the FF indices allows one to derive the combination formulas for the condensed quantities, i.e., the fragment chemical potentials and condensed hardness matrix elements in terms of the relevant AIM charge sensitivities:

$$\mu_A^M = \left(\frac{\partial E^G}{\partial N_A}\right)_{N_{B \neq A}} = \left(\frac{\partial N}{\partial N_A}\right)\left(\frac{\partial E^{AIM}}{\partial N}\right) = \sum_B^M \left(\frac{\partial N_B}{\partial N_A}\right)\left(\frac{\partial E}{\partial N_B}\right)$$

$$= \sum_B^M f^{A,B}(\mu_B)^\dagger, \quad A, B = (X, Y, ..., Z) \tag{130}$$

$$\eta_{A,B}^M = \frac{\partial^2 E^G}{\partial N_A \partial N_B} = \frac{\partial \mu_A^M}{\partial N_B} = \frac{\partial \mu_B^M}{\partial N_A} = \sum_{C,D}^M \frac{\partial N_C}{\partial N_A} \frac{\partial^2 E}{\partial N_C \partial N_D} \frac{\partial N_D}{\partial N_B}$$

$$= \sum_{C,D}^M f^{A,C} \eta^{C,D} (f^{B,D})^\dagger, \quad A, B, C, D = (X, Y, ..., Z) \tag{131}$$

One can also generalize the Berkowitz-Parr formula of Eqs. (16) and (55) into the following form, valid for the constrained equilibrium in M^G:

$$\sigma^{A,B} = -\beta^{A,B} + \sum_{C,D}^M (f^{D,A})^\dagger \sigma_{D,C}^M f^{C,B} \tag{132}$$

where the condensed softness matrix element is

$$\sigma_{A,B}^M = \frac{\partial N_B}{\partial \mu_A} = \sum_C^M \frac{\partial N_C}{\partial \mu_A} \frac{\partial N_B}{\partial N_C} \equiv S^{A,B} 1_B^\dagger = 1_A \sigma^{A,B} 1_B^\dagger \tag{133}$$

3.2 Collective Modes of Molecular Fragments

The concepts of PNM and MEC/REC may be extended into those representing molecular fragments corresponding to a given partitioning of the system in question into mutually closed, interacting subsystems. As we have argued above, such an extension can be of great interest in the theory of chemical reactivity. For example, in gas-phase reactions one is interested in the role played by the charge displacement modes of the A and B reactants in $M^+ = (A|B)$, before the $B \to A$ CT; of interest also are the changes in the PNM/MEC/REC of reactants in M^+ due to the presence of the other reactant, relative to those in the limit of separated reactants.

In the M^+ case, complementary fragments (reactants) are considered to be mutually closed subsystems. One encounters a similar type of partitioning in the physisorption catalytic system consisting of the catalyst ($\mathscr{C}$) and the adsorbate ($\mathscr{A}$): ($\mathscr{A}|\mathscr{C}$). However, in the chemisorption systems ($\mathscr{A}\,\vdots\,\mathscr{C}$), both fragments, adsorbate and substrate, are regarded mutually opened. When two reactants $\mathscr{A}_1$ and $\mathscr{A}_2$ are chemisorbed on a surface, they are both externally open subsystems, in contact with their respective active sites $\mathscr{C}_1$ and $\mathscr{C}_2$. Both adsorbates may be considered as mutually closed, ($\mathscr{C}_1\,\vdots\,\mathscr{A}_1|\mathscr{A}_2\,\vdots\,\mathscr{C}_2$), when referring to their hypothetical state before the direct CT: ($\mathscr{C}_1\,\vdots\,\mathscr{A}_1$) $\to$ ($\mathscr{A}_2\,\vdots\,\mathscr{C}_2$). Clearly, instead of mentally separating the charges of the two active sites from the rest of the surface, one may also consider them as parts of $\mathscr{C}$, open with respect to one another, thus envisaging the two adsorbates as able to exchange electrons indirectly, through the catalyst: ($\mathscr{A}_1\,\vdots\,\mathscr{C}\,\vdots\,\mathscr{A}_2$).

Yet another impotant case of a partitioning of model catalytic systems involves the cluster (C) representations of $\mathscr{C}$, $C = (C_{as}|D)$, where C_{as} is a *small* cluster representing an *active site* (as) of the surface cluster C, while D is a *large* cluster remainder, which electronically conditions the adsorption/rearrangement processes on a surface and provides an environment which moderates the $\mathscr{A} - C_{as}CT$. Clearly, all alternative divisions of a general catalytic system involving $\mathscr{A}$ (or $\mathscr{A}_1$ and $\mathscr{A}_2$), C, and an infinite system remainder, R, can be examined, in order to characterize different aspects (hypothetical intermediate stages) of such elementary adsorption/desorption/rearrangement processes, e.g., $(\mathscr{A}|C_{as}|D|R)$, $(\mathscr{A}|C_{as} \vdots D|R)$, $(\mathscr{A} \vdots C_{as} \vdots D|R)$, etc. The last of these partitionings, with $\mathscr{A}$ and C in equilibrium, could then be used to probe the CT behavior of the $(\mathscr{A} \vdots C)$ system as a whole towards an external agent, e.g., the second adsorbate.

Let us first consider the $M^+ = (A|B)$ system, consisting of an acidic (A) and basic (B) pair of reactants. As explicitly shown elsewhere [9, 15, 20, 25], the presence of the other reactant modifies the respective blocks $\eta^M_{X,Y}$, of the hardness matrix, due to relaxational contributions. As we shall demonstrate in the next section, the relaxed hardness matrix of Eq. (1a) corresponds to the collective relaxational modes $N^{rel} = Nt$, $\eta^{rel} = t^{-1}\eta t$, where the transformation matrix t is defined by the relaxational matrices $T^{(A|B)}$ and $T^{(B|A)}$ (Eqs. (39) and (42)). This transformation and the eigenvectors of the diagonal blocks $\eta^M_{A,A}$ and $\eta^M_{B,B}$,

$$(U^M_X)^\dagger \eta^M_{X,X} U^M_X = h^M_X \text{ (diagonal)}, \quad (U^M_X)^\dagger U^M_X = I_X, \quad X = A, B \quad (134)$$

define new sets of the relaxed PNM of the reactants. Namely, by combining the U^M_X blocks into the corresponding intra-fragment decoupling transformation,

$$\tilde{U}^{rel}_{int} = \begin{pmatrix} U^M_A & 0_{A,B} \\ 0_{B,A} & U^M_B \end{pmatrix} \quad (135)$$

one obtains the relaxed PNM of interacting reactants in terms of the AIM populations of M:

$$d n^{rel}_{int} = d Nt \tilde{U}^{rel}_{int} \equiv d N U^{rel}_{int} \quad (136)$$

which should differ from those of isolated reactants. The latter are defined by the matrix;

$$U_{int} = \begin{pmatrix} U_A & 0_{A,B} \\ 0_{B,A} & U_B \end{pmatrix} \quad (137)$$

where the unitary block U_X diagonalizes the corresponding diagonal block $\eta^{X,X}$ of the rigid AIM hardness matrix of Eq. (1).

Obviously, one may also probe M^+ via the test charge displacements of AIM in one reactant, say on atom a in A. Such a forced, localized displacement $\Delta N_a = 1$ creates, via the minimum energy criterion for the assumed charge separation between the reactants, the collective charge displacement on both reactants, consisting of the internal MEC part in A and the associated REC component in B:

$$\vec{\imath}_a^{M^+} = \vec{n}_{A(a)}^{M^+} + \vec{\mathcal{R}}_{B(a)}^{M^+} \tag{138}$$

where:

$$\vec{n}_{A(a)}^{M^+} = \sum_{a'}^{A} (\bar{a}')_a^{M^+} \vec{e}_{a'}, \quad (\bar{a}')_a^{M^+} \equiv \left(\frac{\partial N_{a'}}{\partial N_a}\right)_{N_A, N_B} = \frac{S_{a', a}^{M^+}}{S_{a, a}^{M^+}} \tag{139}$$

and

$$\vec{\mathcal{R}}_{B(a)}^{M^+} = \sum_{b}^{B} [b]_a^{M^+} \vec{e}_b, \quad [b]_a^{M^+} \equiv \left(\frac{\partial N_b}{\partial N_a}\right)_{N_A, N_B} \tag{140}$$

where $S_{A,A}^{M^+} = (\partial N_A / \partial \mu_A)_{N_A, N_B}^{M^+}$ is the (relaxed) internal softness matrix of A in M^+. The REC components are defined by the relevant relaxational matrix $T^{(A|B)}$ (Eqs. (39) and (42)) and the perturbing MEC displacement $\vec{n}_{A(a)}^{M^+}$:

$$[b]_a^{M^+} = \sum_{a'}^{A} (\bar{a}')_a^{M^+} T_{a', b}^{(A|B)} \tag{141}$$

In yet another hypothetical case in which the primary displaced fragment is in equilibrium with an external reservoir (r), $M_A' = (r|A|B)$, one has to replace the internal MEC, $\vec{n}_{A(a)}^{M^+}$ in Eqs. (138)–(141), by the corresponding external MEC, $\vec{\mathcal{N}}_{A(a)}^{M_A'} = \sum_{a'}^{A} (a')_a^{M_A'} \vec{e}_{a'}$, where $(a')_a^{M_A'} \equiv (\partial N_{a'} / \partial N_a)_{N_B} = \sigma_{a' a}^{M_A'} / \sigma_{a, a}^{M_A'}$, and $\sigma_{A,A}^{M_A'} = (\partial N_A / \partial \mu_A)_{N_B}$ is the (relaxed) external softness matrix of A in M_A'. The resulting expression for the accompanying relaxational adjustment $\vec{\mathcal{R}}_{B(a)}^{M_A'}$ in

$$\vec{\imath}_a^{M_A'} = \vec{\mathcal{N}}_{A(a)}^{M_A'} + \vec{\mathcal{R}}_{B(a)}^{M_A'} \tag{142}$$

is:

$$\vec{\mathcal{R}}_{B(a)}^{M_A'} = \sum_{b}^{B} [b]_a^{M_A'} \vec{e}_b, \quad [b]_a^{M_A'} = \left(\frac{\partial N_b}{\partial N_a}\right)_{N_B} = \sum_{a'}^{A} (a')_a^{M_A'} T_{a', b}^{(A|B)} \tag{143}$$

The above MEC/REC generalizations can also be extended to a general $M = (A|B|C|D|\ldots) \equiv (M_{open}|M_{closed})$ partitioning, including the mutually open fragments in M_{open}, and the mutually closed fragments in M_{closed}. Namely, any unit displacement on an atom in M_{open} will generate the relevant MEC part in this subsystem and the conjugated REC parts in each (internally open) molecular fragment of M_{closed}.

3.3 Relaxed Fragment Transformation of Reactant Hardness Matrix

As we have already remarked in the preceding section, the relaxed hardness matrix of $M = (A|B)$ (Eq. (1a)),

$$\eta^{rel} \equiv \eta + \delta\eta(M) = \begin{pmatrix} \eta_{A,A}^{M} & \eta_{A,B}^{M} \\ \eta_{B,A}^{M} & \eta_{B,B}^{M} \end{pmatrix} \tag{144}$$

represents the transformed (rigid) AIM hardness tensor, η, since it corresponds to the representation of the collective *relaxational modes* (REM), $\hat{\boldsymbol{e}}_M^{rel} = \hat{\boldsymbol{e}}_M \boldsymbol{t}$, defined by the transformation

$$\boldsymbol{t} = \begin{pmatrix} \boldsymbol{I}_A & \boldsymbol{T}^{(A|B)} \\ \boldsymbol{T}^{(B|A)} & \boldsymbol{I}_B \end{pmatrix} = (\boldsymbol{t}_1 | \boldsymbol{t}_2 | \dots | \boldsymbol{t}_m) \tag{145}$$

where $\boldsymbol{T}^{(A|B)}$ and $\boldsymbol{T}^{(B|A)}$ are the relaxational matrices of Eqs. (39) and (42). The corresponding mode population variables and chemical potentials are:

$$\mathrm{d}\boldsymbol{N}^{rel} = \mathrm{d}\boldsymbol{N}\boldsymbol{t} \quad \text{and} \quad \boldsymbol{\mu}^{rel} = \boldsymbol{\mu}\boldsymbol{t} \tag{146}$$

Hence, by transforming the AIM-electronegativity equalization equations, $\mathrm{d}\boldsymbol{\mu} = \mathrm{d}\boldsymbol{N}\boldsymbol{\eta} \equiv -\boldsymbol{\delta N}\boldsymbol{\eta}$, one obtains their REM representation:

$$\mathrm{d}\boldsymbol{\mu}^{rel} = (\mathrm{d}\boldsymbol{N}\boldsymbol{t})(\boldsymbol{t}^{-1}\boldsymbol{\eta}\boldsymbol{t}) \equiv \mathrm{d}\boldsymbol{N}^{rel}\boldsymbol{\eta}^{rel} \tag{147}$$

which identifies $\boldsymbol{\eta}^{rel}$ as the similarity transformed $\boldsymbol{\eta}$.

We would like to emphasize that the REM basis vectors, $\hat{\boldsymbol{e}}_M^{rel} = \{\hat{\boldsymbol{e}}_k^{rel} = \hat{\boldsymbol{e}}_M \boldsymbol{t}_k\}$, are not orthonormal, since their metric

$$(\hat{\boldsymbol{e}}_M^{rel})^\dagger \hat{\boldsymbol{e}}_M^{rel} = \boldsymbol{t}^\dagger \hat{\boldsymbol{e}}_M^\dagger \hat{\boldsymbol{e}}_M \boldsymbol{t} = \boldsymbol{t}^\dagger \boldsymbol{t} \neq \boldsymbol{I} \tag{148}$$

One can express the corresponding blocks of the inverse transformation $\boldsymbol{t}^{-1} \equiv \{\boldsymbol{t}_{X,Y}^{-1}, (X, Y) = (A, B)\}$ in terms of the corresponding blocks of $\boldsymbol{t}$:

$$\boldsymbol{t}_{A,A}^{-1} = [\boldsymbol{I}_A - \boldsymbol{T}^{(A|B)}\boldsymbol{T}^{(B|A)}]^{-1}, \quad \boldsymbol{t}_{B,B}^{-1} = [\boldsymbol{I}_B - \boldsymbol{T}^{(B|A)}\boldsymbol{T}^{(A|B)}]^{-1}$$

$$\boldsymbol{t}_{A,B}^{-1} = -[\boldsymbol{I}_A - \boldsymbol{T}^{(A|B)}\boldsymbol{T}^{(B|A)}]^{-1}\boldsymbol{T}^{(A|B)},$$

$$\boldsymbol{t}_{B,A}^{-1} = -[\boldsymbol{I}_B - \boldsymbol{T}^{(B|A)}\boldsymbol{T}^{(A|B)}]^{-1}\boldsymbol{T}^{(B|A)} \tag{149}$$

The power expansion of the matrix $(\boldsymbol{I} - \boldsymbol{x})^{-1}$,

$$(\boldsymbol{I} - \boldsymbol{x})^{-1} = \sum_{i=0}^{\infty} \boldsymbol{x}^i, \tag{150}$$

then allows one to express the relaxational corrections $\boldsymbol{\delta\eta}_{X,X}(M)$ and $\boldsymbol{\delta\eta}_{X,Y}(M)$, explicitly in powers of the relevant blocks of $\boldsymbol{\eta}$ and $\boldsymbol{t}$.

The transformation of Eq. (147) gives:

$$\boldsymbol{\eta}_{A,A}^M = \boldsymbol{t}_{A,A}^{-1}[\boldsymbol{\eta}^{A,A} + \boldsymbol{\eta}^{A,B}\boldsymbol{T}^{(B|A)}] + \boldsymbol{t}_{A,B}^{-1}[\boldsymbol{\eta}^{B,A} + \boldsymbol{\eta}^{B,B}\boldsymbol{T}^{(B|A)}]$$

$$\boldsymbol{\eta}_{A,B}^M = \boldsymbol{t}_{A,A}^{-1}[\boldsymbol{\eta}^{A,A}\boldsymbol{T}^{(A|B)} + \boldsymbol{\eta}^{A,B}] + \boldsymbol{t}_{A,B}^{-1}[\boldsymbol{\eta}^{B,A}\boldsymbol{T}^{(A|B)} + \boldsymbol{\eta}^{B,B}] \tag{151}$$

and the remaining expressions immediately follow from exchanging the fragment symbols in Eq. (151). The power expansion (Eq. (150)) of $(\boldsymbol{I}_A - \boldsymbol{T}^{(A|B)}\boldsymbol{T}^{(B|A)})^{-1} \equiv (\boldsymbol{I}_A - \boldsymbol{x}_A)^{-1}$ gives:

$$\boldsymbol{t}_{A,A}^{-1} \equiv \sum_{i=0}^{\infty} \boldsymbol{x}_A^i \quad \text{and} \quad \boldsymbol{t}_{A,B}^{-1} \equiv -\left[\sum_{i=0}^{\infty} \boldsymbol{x}_A^i\right]\boldsymbol{T}^{(A|B)} \tag{152}$$

and hence, to the second-order in $\mathbf{T}^{(X|Y)}$:

$$\eta_{A,A}^M = \eta^{A,A} + [\eta^{A,B}\mathbf{T}^{(B|A)} - \mathbf{T}^{(A|B)}\eta^{B,A} + x_A\eta^{A,A}$$
$$- \mathbf{T}^{(A|B)}\eta^{B,B}\mathbf{T}^{(B|A)} + \dots] \tag{153}$$
$$\equiv \eta^{A,A} + \delta\eta_{A,A}(M)$$

$$\eta_{A,B}^M = \eta^{A,B} + [\eta^{A,A}\mathbf{T}^{(A|B)} - \mathbf{T}^{(A|B)}\eta^{B,B} + x_A\eta^{A,B}$$
$$- \mathbf{T}^{(A|B)}\eta^{B,A}\mathbf{T}^{(A|B)} + \dots] \tag{154}$$
$$\equiv \eta^{A,B} + \delta\eta_{A,B}(M)$$

We would like to emphasize that the relaxed hardness data can also be obtained by the single matrix inversion procedure of Sect. 3.1. Namely, in order to obtain the diagonal block $\eta_{A,A}^M$, one assumes the partitioning $M_A^{rel} = (1|2|\dots|n|B)$, in which all n AIM in A are mutually closed, while the remaining $m - n$ constituent atoms belonging to B can exchange electrons freely. Equation (128) then gives $\{\eta_{a,a'}\} = \eta_{A,A}^M$, and the relaxed off-diagonal parameters $\eta_{a,B}^M$ that measure the effective coupling hardness between $a \in A$ and the relaxed B as a whole.

4 Donor-Acceptor Reactive Systems

4.1 *In Situ* Quantities

In this section we briefly summarize the *in situ* electron-population/chem-ical-potential derivatives along the N-restricted CT process, $dN = 0$ or $dN_A = - dN_B = N_{CT}$, in a general system $M = A - - - B$, consisting of an acidic (A) and basic (B) pair of reactants [25]. We shall adopt the AIM resolution of the preceding sections. The *in situ* derivatives involve the differentiation with respect to the amount of CT, N_{CT}.

As we have already observed before, the closed M constraint implies trivial dependencies, $f_A = \partial N_A/\partial N_{CT} = 1$ and $f_B = \partial N_B/\partial N_{CT} = - 1$, which enter the *in situ* derivatives. In the hypothetical state $M^+ = (A|B)$, when both react-ants are mutually closed but fully relaxed internally, one obtains the *in situ* chemical-potential difference,

$$\mu_{CT}^+ = \frac{\partial E(A|B)}{\partial N_A}f_A + \frac{\partial E(A|B)}{\partial N_B}f_B = \mu_A^M - \mu_B^M \tag{155}$$

while the effective *in situ* hardness,

$$\eta_{CT} = \eta_A^M + \eta_B^M \tag{156}$$

is the sum of the *in situ* hardnesses of reactants, $\eta_A^M = \eta_{A,A}^M - \eta_{A,B}^M$ and $\eta_B^M = \eta_{B,B}^M - \eta_{B,A}^M$. The condition for a vanishing inter-reactant CT gradient, $g(N_{CT}) = 0$, yields the equilibrium value of N_{CT}, which equalizes the chemical potentials of the mutually open reactants in $M^* = (A|B)$.

Consider next the *in situ* CT softnesses, which involve derivatives with respect to the fragment chemical potentials as well as related quantities, e.g., the FF indices. The global CT softness is immediately given by the inverse of η_{CT},

$$S_{CT} = \partial N_{CT}/\partial \mu_{CT} = 1/\eta_{CT} \tag{157}$$

The condensed reactant FF indices, f_A and f_B, also represent the *in situ* quantities, and as such must sum to zero, $f_A + f_B = 0$, in accordance with the fixed global number of electrons in the system.

For the mixed derivatives that represent the *in situ* AIM hardnesses in AB, e.g., for constituent atoms in A, one obtains:

$$
\eta_A^{CT} \equiv \frac{\partial^2 E(A|B)}{\partial N_A \partial N_{CT}} = \frac{\partial \mu_{CT}}{\partial N_A} = \frac{\partial \mu_A}{\partial N_{CT}} = f_A \left(\frac{\partial \mu_A}{\partial N_A} \right)_{N_B} + f_B \left(\frac{\partial \mu_A}{\partial N_B} \right)_{N_A}
$$

$$
= \left(\frac{\partial N}{\partial N_A} \right)_{N_B} \left(\frac{\partial \mu_A}{\partial N} \right)_{N_B} - \left(\frac{\partial N}{\partial N_B} \right)_{N_B} \left(\frac{\partial \mu_A}{\partial N} \right)_{N_A} \tag{158}
$$

$$
= (f^{A,A} - f^{B,A})\eta_{A,A}^M - (f^{B,B} - f^{A,B})\eta_{B,A}^M \equiv f_A^{CT}\eta_{A,A}^M - f_B^{CT}\eta_{B,A}^M
$$

where $f^{X,Y}$ and $\eta_{X,Y}^M$, $(X, Y) = (A, B)$ represent the relaxed AIM FF indices and hardnesses, respectively.

Finally, the global *in situ* hardness can also be related to the above AIM quantities by

$$
\eta_{CT} = f_A \left(\frac{\partial \mu_{CT}}{\partial N_A} \right)_{N_B} + f_B \left(\frac{\partial \mu_{CT}}{\partial N_A} \right)_{N_A} = \left(\frac{\partial N}{\partial N_A} \right)_{N_B} \left(\frac{\partial \mu_{CT}}{\partial N} \right)
$$

$$
- \left(\frac{\partial N}{\partial N_B} \right)_{N_A} \left(\frac{\partial \mu_{CT}}{\partial N} \right) = f_A^{CT}(\eta_A^{CT})^\dagger - f_B^{CT}(\eta_B^{CT})^\dagger \tag{159}
$$

The above CT FF indices (Eq. 158) define the partitioning of the global CT softness into the corresponding atomic contributions, e.g.:

$$
S_A^{CT} = \frac{\partial N_{CT}}{\partial \mu_A} = \frac{\partial N_A}{\partial \mu_{CT}} = \frac{\partial N_{CT}}{\partial \mu_{CT}} \left(\frac{\partial N_A}{\partial N_{CT}} \right)_N = S^{CT} f_A^{CT} \tag{160}
$$

The combination formula for μ_{CT} in terms of the relaxed chemical potentials of constituent atoms,

$$
\mu_A^M = \left(\frac{\partial E(A|B)}{\partial N_A} \right)_{N_B} = \left(\frac{\partial E(A|B)}{\partial N_A} \right)_{N_B} + \frac{\partial N_B}{\partial N_A} \left(\frac{\partial E(A|B)}{\partial N_B} \right)_{N_A} \tag{161}
$$

$$
= \mu_A^{rgd} + \mu_B^{rgd}(T^{(A|B)})^\dagger \equiv \mu_A^{rgd} + \delta\mu_A(M)
$$

$$
\mu_{CT} = \frac{\partial N_A}{\partial N_{CT}} \left(\frac{\partial E(A|B)}{\partial N_A} \right)_{N_B} = f_A^{CT}(\mu_A^M)^\dagger \tag{162}
$$

4.2 Interaction Energy and Its Partitioning

In this section we examine—within the local resolution—the quadratic interaction energy in a general reactive system, $M = A - \overset{Q}{\underset{}{\_\_}} - B$, consisting of an electron donor [basic (B)] and an electron acceptor [acidic (A)] reactants at their current mutual separation and orientation, Q [3, 8, 9, 25, 39–41]. We follow the standard partitioning scheme of Morokuma [42], and previous DFT approaches to *electronegativity equalization*, and the corresponding definition of molecular fragments [3, 8, 9, 25, 43], e.g., AIM. This approach considers the following stages of the reactant electron distribution in M:

(i) the separated (infinitely distant) reactants: $M^0(\infty) = A^0 + B^0$;
(ii) the polarized mutually closed reactants at Q: $M^+(Q) = (A^+ | B^+)$;
(iii) the mutually open reactants at Q: $M^*(Q) = (A^* \vdots B^*)$.

The second stage corresponds to the *intra-reactant equilibrium* with $\mu_A^+ \neq \mu_B^+$, while the third stage represents the *global equilibrium* in M with fully equalized chemical potential (electronegativity), $\mu_A^* = \mu_B^* = \mu_M$, which is attained after the inter-reactant $B \rightarrow A$ CT. One could also formally consider an additional stage (iv), intermediate between (i) and (ii), of "rigid", closed (interacting) reactants at Q, $M^0(Q) = (A^0 | B^0)$ in which no electronegativity equalization within reactants has taken place.

The reactants in $M^*(Q)$ are displaced relative to those in $M^0(\infty)$ in both the external potential, $dv = [dv_A, dv_B] \approx [\phi_B, \phi_A]$ and the electron populations, $dN = [dN_A, dN_B] = [N_{CT}, -N_{CT}]$; here $\phi_X(r) = v_X(r) + \int dr' \rho_X^0(r')/|r - r'|$ is the electrostatic (Hartree) potential around $X = A, B$ (defined with respect to the electron charge), where $v_X(r) = -\sum_i^X Z_i/|r - R_i|$. The total interaction energy between A and B, $E_{AB} = E_{AB}^e + V_{AB}^{nn}$, consists of the corresponding electronic contribution, E_{AB}^e, and a trivial nuclear repulsion part, V_{AB}^{nn}. One observes that the quadratic approximation to E_{AB}^e,

$$E_{AB}^e = dN\left(\mu^0 + \frac{1}{2}d\mu\right)^\dagger + \int dv(r)[\rho^0(r) + \tfrac{1}{2}d\rho(r)]^\dagger dr \tag{163}$$

where $\rho^0 = [\rho_A^0, \rho_B^0]$ and $\mu^0 = [\mu_A^0, \mu_B^0]$, involves only the first-order responses $d\mu$ and $d\rho$, conjugated to the perturbations dN and dv, respectively. These responses can be partitioned conveniently into polarizational parts (due to dv, at $dN = 0$) and the remaining CT contributions (due to N_{CT}, at $dv \neq 0$):

$$d\mu = d\mu^+(dv) + d\mu^*(N_{CT}) = [d\mu_A, d\mu_B] \tag{164}$$

$$d\rho = d\rho^+(dv) + d\rho^*(N_{CT}) = [d\rho_A, d\rho_B] \tag{165}$$

First, let us consider the polarizational stage, $M^+(Q)$. The corresponding changes in μ and ρ are:

$$d\mu^+ = \int dv(r)\left(\frac{\delta\mu}{\delta v(r)}\right)_N dr = \int dv(r) f^\dagger(r) dr \tag{166}$$

$$d\rho^+(r) = \int d\,v(r')\left(\frac{\delta\rho(r)}{\delta v(r')}\right)_N dr' \equiv \int d\,v(r')\,\beta(r',r)\,d\,r' \tag{167}$$

The local FF matrix in Eq. (166),

$$\mathbf{f}(r) = \left\{f_{Y,X}(r) = \frac{\delta\mu_Y}{\delta v_X(r)} = \frac{\partial\rho_X(r)}{\partial N_Y} = \frac{\delta^2 E_{AB}^e}{\delta N_Y\,\delta v_X(r)}\right\},$$

$$(X, Y) = (A, B) \tag{168}$$

includes the diagonal elements, $f_{X,X}(r) = \partial\rho_X(r)/\partial N_X$, normalized to $\int f_{X,X}(r)\,dr = \partial N_X/\partial N_X = 1$, and the off-diagonal (relaxational) elements, $f_{X,Y}(r) = \partial\rho_Y(r)/\partial N_X$, normalized to $\int f_{X,Y}(r)\,dr = \partial N_Y/\partial N_X = 0$. For a given reactant, say A:

$$d\mu_A^+ = \int[dv_A(r)\,f_{A,A}(r) + dv_B(r)\,f_{A,B}(r)]\,dr, \quad \text{etc.} \tag{169}$$

The density polarization matrix kernel of Eq. (167),

$$\beta(r',r) = \left\{\beta_{X,Y}(r',r) = \frac{\delta\rho_Y(r)}{\delta v_X(r')} = \frac{\delta^2 E_{AB}^e}{\delta v_X(r')\,\delta v_Y(r)}\right\}, \quad (X, Y) = (A, B) \tag{170}$$

also involves the coupling (off-diagonal) response kernels, which contribute to the first-order change in the density of polarized reactants

$$d\rho_A^+(r) = \int[dv_A(r')\,\beta_{A,A}(r',r) + dv_B(r')\,\beta_{B,A}(r',r)]\,dr', \quad \text{etc.} \tag{171}$$

The polarizational shifts in the chemical potentials of the reactant (Eqs. (166) and (169)) determine $\mu_{CT} = \mu_{CT}^0 + d\mu_A^M - d\mu_B^M$, which, together with η_{CT}, determines both N_{CT} and E_{CT}, as well as the CT displacements in μ and ρ:

$$d\mu_X^* = \frac{\partial\mu_X}{\partial N_{CT}} N_{CT} = \eta_X^M N_{CT}, \quad X = A, B \tag{172}$$

$$d\rho^*(r) = \frac{\partial\rho(r)}{\partial N_{CT}} N_{CT} = f^{CT}(r)\,N_{CT} \tag{173}$$

where the two components of $f^{CT}(r)$ are:

$$f_A^{CT}(r) = f_{A,A}(r) - f_{B,A}(r) \quad \text{and} \quad f_B^{CT}(r) = f_{B,B}(r) - f_{A,B}(r) \tag{174}$$

In this approach, the external potential displacements that are responsible for a transition from stage (i) to stage (ii) create conditions for the subsequent CT effects, in the spirit of the Born–Oppenheimer approximation. Clearly, the consistent second-order Taylor expansion at $M^0(\infty)$ does not involve the coupling hardness $\eta_{A,B}$ and the off-diagonal response quantities of Eqs. (168) and (170), which vanish identically for infinitely separated reactants. However, since the interaction at Q modifies both the chemical potential difference and the second-order charge sensitivities (non-vanishing off-diagonal elements), one should obtain a more realistic estimate of the interaction energy by considering the CT stage at the finite separation between the reactants [8, 25]. In this approach the overall interaction energy is obtained as the sum of the energy differences between the above three stages of interaction:

$$E_{AB}^e = [E_{AB}^e(ii) - E_{AB}^e(i)] + [E_{AB}^e(iii) - E_{AB}^e(ii)] \tag{175}$$

$$\equiv E_{AB}^+(d\mathbf{v}, N_{CT} = 0) + E_{AB}^*(d\mathbf{v}, N_{CT})$$

In the quadratic approximation, one uses the diagonal second-order derivatives of the isolated reactants with respect to the external potential perturbations at $M^0(\infty)$ to determine properties of $M^+(Q)$; one then calculates the charge transfer energy, which explicitly involves the off-diagonal charge sensitivities, which are non-vanishing at Q from the relevant second-order Taylor expansion in N_{CT} at $M^+(Q)$. Again, the diagonal elements in Eqs. 168 and 170 can be approximated by the corresponding quantities for isolated reactants; additionally, one has to determine the off-diagonal charge sensitivities, which can be approximated by those at stage (iv), in $M^0(Q)$.

For interpretative purposes, it is important to partition the overall electronic interaction energy $E_{AB}^e(d\mathbf{v}, N_{CT})$, into contributions attributed to reactants A (acidic) and B (basic) in the closed $M = A - B$, or to their constituent fragments. Following Eq. (175), we consider separately the (electrostatic + polarization) $-$ energy, $E_{AB}^+(d\mathbf{v}, N_{CT} = 0) = E_{ES} + E_P = E(M^+) - E(M^0)$, and the CT energy:

$$E_{CT} = E_{AB}^*(d\mathbf{v}, N_{CT}) = E(M^*) - E(M^+) = \tfrac{1}{2}\mu_{CT}^+ N_{CT} \tag{176}$$

where:

$$N_{CT} = -\mu_{CT}^+/\eta_{CT} \tag{177}$$

The first part can be naturally partitioned into the reactant contributions:

$$E_{AB}^+ = \sum_X^{A,B} \int d\mathbf{v}_X(\mathbf{r})[\rho_X^0(\mathbf{r}) + \tfrac{1}{2}d\rho_X^+(\mathbf{r})]\,d\mathbf{r} \equiv E_A^+ + E_B^+ \tag{178}$$

The CT energy is also naturally partitioned into the reactants contributions:

$$E_{CT} = \sum_X^{A,B} \left\{ dN_X[\mu_X^0 + \tfrac{1}{2}(d\mu_X^M + d\mu_X^*)] + \tfrac{1}{2}\int d\mathbf{v}_X(\mathbf{r})\,d\rho_X^*(\mathbf{r})\,d\mathbf{r} \right\}$$

$$\equiv E_A^* + E_B^* \tag{179}$$

Consider now the phenomenological AIM discretization of the above energy expressions, in which $d\mathbf{N}_X = \{dN_x\}$ and $d\mathbf{v}_X = \{dv_x\}$ replace $d\rho_X(\mathbf{r})$ and $d\mathbf{v}_X(\mathbf{r})$, respectively. In this resolution:

$$dN_X^+ = \sum_Y^{A,B} d\mathbf{v}_Y \beta^{Y,X}, \quad X = A, B \quad \text{or} \quad dN_X^+ = d\mathbf{v}\beta \tag{180}$$

$$d\mu_X^+ = \sum_Y^{A,B} d\mathbf{v}_Y f^{X,Y}, \quad X = A, B \quad \text{or} \quad d\mu^+ = d\mathbf{v}f^\dagger \tag{181}$$

where, in short notation, $d\mathbf{N} = (d\mathbf{N}_A, d\mathbf{N}_B)$, $d\mathbf{v} = (d\mathbf{v}_A, d\mathbf{v}_B)$, $d\boldsymbol{\mu}^+ = (d\boldsymbol{\mu}_A^+, d\boldsymbol{\mu}_B^+)$, $\boldsymbol{\beta} = \partial\mathbf{N}/\partial\mathbf{v}$, and $\mathbf{f}^{\dagger} = \partial\boldsymbol{\mu}^+/\partial\mathbf{v}$. Similarly, the CT displacements are:

$$d\mathbf{N}_X^* = f_X^{CT} N_{CT}, \quad X = A, B \quad \text{or} \quad d\mathbf{N}^* = \mathbf{f}^{CT} N_{CT} \tag{182}$$

$$d\mu_X^* = \eta_X^M N_{CT}, \quad X = A, B \quad \text{or} \quad d\boldsymbol{\mu}^* = \boldsymbol{\eta}^M N_{CT} \tag{183}$$

where:

$$\eta_X^M = \eta_{X,X}^M - \eta_{X,Y}^M \, (X \neq Y), \quad \boldsymbol{\eta}^M = (\eta_A^M, \eta_B^M) = \partial\boldsymbol{\mu}/\partial N_{CT},$$

$$f_X^{CT} = f^{X,X} - f^{Y,X} \, (X \neq Y) = \partial\mathbf{N}_X/\partial N_{CT}, \quad \mathbf{f}^{CT} = (f_A^{CT}, f_B^{CT})$$

Thus, in the AIM discretization,

$$E_X^+ = (\mathbf{N}_X^0 + \tfrac{1}{2} d\mathbf{N}_X^+)(d\mathbf{v}^X)^{\dagger} \tag{184}$$

$$E_X^* = [\mu_X^0 + \tfrac{1}{2}(d\mu_X^M + d\mu_X^*)] \, d\mathbf{N}_X + \tfrac{1}{2} d\mathbf{N}_X^*(d\mathbf{v}_X)^{\dagger}, \quad X = A, B \tag{185}$$

where $d\mathbf{N}_A = N_{CT}$ and $d\mathbf{N}_B = -N_{CT}$. The expressions above provide a unique partitioning of E_X^+ and E_X^* into the contributions of the constituent atoms.

One can also partition the external CT energy of electron exchange between any given molecular system M and the reservoir r. Let

$$\mu'_{CT} \equiv \mu_M - \mu_r, \quad \eta'_{CT} \equiv \eta_M - \eta_{M,r} > 0 \tag{186}$$

where μ_r stands for the chemical potential of the reservoir and $\eta_{M,r}$ represents a weak-coupling hardness between the system and the reservoir. We have neglected the reservoir diagonal hardness which vanishes by assumption (infinitely soft, macroscopic reservoir). Then, $N'_{CT} > 0$ for $\mu'_{CT} < 0$ (inflow of electrons to M) and $N'_{CT} < 0$ for $\mu'_{CT} > 0$ (outflow of electrons from M), and the CT stabilization energy can be partitioned into the relevant AIM contributions by partitioning N'_{CT} into changes in the atomic populations:

$$E'_{CT} = \frac{1}{2}\mu'_{CT} N'_{CT} = (\textstyle\sum_i^M f_i) E'_{CT} \equiv \textstyle\sum_i^M \varepsilon'_{CT,i} \tag{187}$$

For the molecular-fragment partitioning in the global equilibrium, $M = (X\mathbin{|}Y\mathbin{|}\ldots)$, one calculates the associated group contributions to E'_{CT} by summation over all constituent atoms:

$$E'_{CT} = \textstyle\sum_X^M \sum_x^X \varepsilon'_{CT,x} \equiv \textstyle\sum_X^M E'_{CT,X} \tag{188}$$

in which $E'_{CT,X} = f_X^M E'_{CT}$, due to the additivity of the AIM FF indices.

Such a FF partitioning of N_{CT} can also be used to partition E_{CT}. Beside the resulting B- and A-resolved divisions of E_{CT},

$$E_{CT} = (\textstyle\sum_a^A f_a^A) E_{CT} \equiv \textstyle\sum_a^A \in_{CT,a} = (\textstyle\sum_b^B f_b^B) E_{CT} \equiv \textstyle\sum_b^B \in_{CT,b}, \tag{189}$$

one can adopt a deeper division [9], which also involves the chemical potential difference $\mu_{CT} = \mu_A^M - \mu_B^M$. Since $N_{CT} = -d\mathbf{N}_B = -\sum_b^B d\mathbf{N}_b$, $d\mathbf{N}_b = f_b^B d\mathbf{N}_B$, and $\mu_B^M = \sum_b^B \mu_b^+ f_b^B$, the following B-resolved expression is obtained for E_{CT}, in which one examines interactions between A as a whole and the AIM local sites (b) in **B**:

$$E_{CT} = \frac{1}{2}\sum_b^B (\mu_b^+ - \mu_A^M)\,dN_b \equiv \sum_b^B \varepsilon_b^{CT} \tag{190}$$

where ε_b^{CT} is the corresponding contribution from the A $-$ b charge flow. The complementary *A-resolved* partitioning, when the intention is to interpret the contributions to interactions between B (treated as a whole) and the individual local sites (a) in A, follows from obvious relations: $N_{CT} = dN_A = \sum_a^A dN_a$, $dN_a = f_a^A dN_A$, $\mu_A^M = \sum_a^A \mu_a^+ f_a^A$

$$E_{CT} = \frac{1}{2}\sum_a^A (\mu_a^+ - \mu_B^M)\,dN_a \equiv \sum_a^A \varepsilon_a^{CT} \tag{191}$$

The deepest partitioning of the AIM resolution is the *(A, B)-resolved* division:

$$E_{CT} = \frac{1}{2}(\mu_B^M\,dN_B + \mu_A^M\,dN_A) = \frac{1}{2}\Big(\sum_b^B \mu_b^+\,dN_b + \sum_a^A \mu_a^+\,dN_a\Big) \tag{192}$$

$$\equiv \sum_b^B \varepsilon_b + \sum_a^A \varepsilon_a \equiv E_B + E_A$$

where the first sum represents the *base charge withdrawal energy* (E_B, positive) and the second sum combines the AIM contributions to the *acid charge inflow* (E_A, negative).

We would like to point out that all CT AIM contributions ε_b^{CT} are stabilizing (negative) since, e.g., for $\mu_b^+ > \mu_A^M$, one obtains by the electronegativity-equalization principle $dN_b < 0$; similarly, $\mu_b^+ < \mu_A^M$ implies $dN_b > 0$. The same electronegativity-equalization argument shows that also all ε_a^{CT} contributions are negative. Thus, the A- and B-resolutions partition the overall CT stabilization energy into stabilizing reactant-atom contributions, while the (A, B)-resolution views the CT process as the consecutive change involving an activation of the basic reactant (removal of N_{CT} electrons from B) followed by charge stabilization in the acidic reactant (adding of N_{CT} electrons to A).

Finally, we would like to comment on the corresponding collective-mode partitionings of the CT energies, E_{CT}^r and E_{CT}. Clearly, the CT-projected w_α indices replace the f_i indices of the AIM-representation in the PNM-representation, e.g.,

$$E_{CT}^r = \Big(\sum_\alpha^M w_\alpha\Big) E_{CT}^r \equiv \sum_\alpha^M \bar\varepsilon_{CT,\alpha}^r, \quad \text{etc.} \tag{193}$$

The E_{CT} energy in the closed A $-$ B system can be partitioned similarly into reactant PNM contributions; in this reference frame $N_{CT} = dN_A = \sum_\alpha^A d\bar n_\alpha = -dN_B = -\sum_\beta^B d\bar n_\beta$, $d\bar n_\alpha = w_\alpha dN_A$, $d\bar n_\beta = w_\beta dN_B$, $\mu_A^M = \sum_\alpha^A \bar\xi_\alpha^+ w_\alpha$, $\mu_B^M = \sum_\beta^B \bar\xi_\beta^+ w_\beta$, etc. Thus, E_{CT} can be expressed in various ways, as:

$$E_{CT} = \frac{1}{2}\sum_\beta^B (\bar\xi_\beta^+ - \mu_A^M)\,d\bar n_\beta \equiv \sum_\beta^B \bar\varepsilon_\beta^{CT} \tag{194}$$

$$E_{CT} = \frac{1}{2} \sum_{\alpha}^{A} (\bar{\xi}_{\alpha}^{+} - \mu_{B}^{M}) d\bar{n}_{\alpha} \equiv \sum_{\alpha}^{A} \bar{\varepsilon}_{\alpha}^{CT} \tag{195}$$

$$E_{CT} = \frac{1}{2} (\sum_{\beta}^{B} \bar{\xi}_{\beta}^{+} d\bar{n}_{\beta} + \sum_{\alpha}^{A} \bar{\xi}_{\alpha}^{+} d\bar{n}_{\alpha}) \equiv \sum_{\beta}^{B} \bar{\varepsilon}_{\beta} + \sum_{\alpha}^{A} \bar{\varepsilon}_{\alpha} \equiv E_{B} + E_{A} \tag{196}$$

These equations closely follow the corresponding expressions (Eqs. 190–192) of the AIM representation.

4.3 Partitioning of the Fukui Function Indices

The basic charge-response quantities of a given open molecular system M, the AIM FF indices, $f_i = \partial N_i / \partial N$, can be partitioned into the respective collective mode contributions. In the reactive system $M^* = (A^* \vert B^*)$, one is also interested in the partitioning of the collective mode terms into the reactant contributions.

Consider the general collective electron population displacements $d\boldsymbol{p}$ (p-modes):

$$d\boldsymbol{p} = d\boldsymbol{N}\boldsymbol{C} \tag{197}$$

defined by the respective columns of the linear transformation matrix $\boldsymbol{C} = (\boldsymbol{C}_1 | \ldots | \boldsymbol{C}_{\alpha} | \ldots | \boldsymbol{C}_m) = \partial \boldsymbol{p} / \partial \boldsymbol{N}$. Let a and b denote representative constituent AIM of A and B, respectively.

A straightforward chain-rule transformation of the AIM FF index f_i gives:

$$f_i = \sum_{\alpha} \frac{\partial p_{\alpha}}{\partial N} \frac{\partial N_i}{\partial p_{\alpha}} \equiv \sum_{\alpha} F_{\alpha}^{(p)} C_{i,\alpha} \equiv \sum_{\alpha} w_{i,\alpha}^{(p)}$$

$$= \sum_{\alpha} C_{i,\alpha} \left(\frac{\partial N_A}{\partial N} \frac{\partial p_{\alpha}}{\partial N_A} + \frac{\partial N_B}{\partial N} \frac{\partial p_{\alpha}}{\partial N_B} \right) \equiv \sum_{\alpha} C_{i,\alpha} (f_A^M F_{\alpha}^{A(p)} + f_B^M F_{\alpha}^{B(p)})$$

$$\equiv \sum_{\alpha} (f_A^M w_{i,\alpha}^{A(p)} + f_B^M w_{j,\alpha}^{B(p)}) \equiv \sum_{\alpha} (W_{i,\alpha}^{A(p)} + W_{i,\alpha}^{B(p)}) \tag{198}$$

where the condensed reactant FF indices f_A^M and f_B^M determine the overall distribution of the external CT among reactants. The above partitioning first divides f_i into the p-mode contributions, $\{w_{i,\alpha}^{(p)}\}$, satisfying the usual normalization,

$$\sum_{\alpha} \left(\sum_{i} w_{i,\alpha}^{(p)} \right) \equiv \sum_{\alpha} w_{\alpha}^{(p)} = 1 \tag{199}$$

and then separates the components reflecting the mode participation in changing the overall electron populations of reactants. These reactant-resolved contributions, the f_X^M weighted $\{W_{i,\alpha}^{X(p)}\}$ and unweighted $\{w_{i,\alpha}^{X(p)}\}$, obey the following global summation rules:

$$\sum_\alpha \sum_i w_{i,\alpha}^{X(p)} = \sum_\alpha \frac{\partial p_\alpha}{\partial N_X} \frac{\partial N}{\partial p_\alpha} = \frac{\partial N}{\partial N_X} = 1 \tag{200}$$

$$\sum_\alpha \sum_i W_{i,\alpha}^{X(p)} = f_X^M \frac{\partial N}{\partial N_X} = f_X^M, \quad X = (A, B) \tag{201}$$

It is also of interest to divide the $w_\alpha^{(p)}$ quantities of Eq. (199) into reactant contributions, defined by the corresponding partial summations over constituent atoms of each reactant:

$$w_\alpha^{(p)} = \sum_a^A w_{a,\alpha}^{(p)} + \sum_b^B w_{b,\alpha}^{(p)} = w_\alpha^{A(p)} + w_\alpha^{B(p)} \tag{202}$$

The collective mode FF index,

$$F_\alpha^{A(p)} = \frac{\partial p_\alpha}{\partial N_A} = \sum_i \frac{\partial N_i}{\partial N_A} \frac{\partial p_\alpha}{\partial N_i} = \sum_a^A f_a^{A,A} C_{a,\alpha} + \sum_b^B f_b^{A,B} C_{b,\alpha} \tag{203}$$

$$\equiv F_\alpha^{A,A(p)} + F_\alpha^{A,B(p)}$$

includes both the diagonal, $f_a^{A,A} \equiv \partial N_a/\partial N_A$, and off-diagonal, $f_b^{A,B} \equiv \partial N_b/\partial N_A$ contributions. Thus, the $F_\alpha^{A(p)}$ index can be partitioned into the diagonal reactant contribution, $F_\alpha^{X,X(p)} = \sum_x^X (\partial N_x/\partial N_X)(\partial p_\alpha/\partial N_x)$, that accounts for the charge reorganization inside X, and the associated off-diagonal part, $F_\alpha^{X,Y(p)} = \sum_y^Y (\partial N_y/\partial N_X)(\partial p_\alpha/\partial N_y)$, $Y \neq X$, due to the equilibrium changes of AIM populations in the other reactant.

The above partitioning applies when M in $M' = (r{:}A{:}B)$ changes the global number of electrons, $dN = dN_A + dN_B \neq 0$. e.g., in catalytic system, when A = adsorbate, B = active site, and the reservoir r = surface reminder. When N is constrained in $M^* = (A{:}B)$, i.e., when $dN_A = -dN_B = N_{CT}$, the reactant FF indices are fixed: $(f_A^M)_N = f_A = 1$ and $(f_A^M)_N = f_B = -1$. For such $N = $ const., internal CT processes, the AIM FF indices are replaced by the corresponding *in situ* indices. Their collective mode partitioning gives:

$$f_x^{CT} = \left(\frac{\partial N_x}{\partial N_{CT}}\right)_N = f_x^{X,X} - f_x^{Y,X} = \sum_\alpha \left[\frac{\partial p_\alpha}{\partial N_X} - \frac{\partial p_\alpha}{\partial N_Y}\right]\frac{\partial N_x}{\partial p_\alpha}$$

$$= \sum_\alpha [F_\alpha^{X(p)} - F_\alpha^{Y(p)}] C_{x,\alpha} = \sum_\alpha (w_{x,\alpha}^{X(p)} - w_{x,\alpha}^{Y(p)}) \equiv \sum_\alpha \omega_{x,\alpha}^{CT(p)} \tag{204}$$

$$(X \neq Y) \in (A, B), \quad x \in X$$

This equation separates the diagonal p-mode CT contributions $\{w_{a,\alpha}^{A(p)}\}$ and $\{w_{b,\alpha}^{B(p)}\}$, from off-diagonal ones, $\{w_{a,\alpha}^{B(p)}\}$ and $\{w_{b,\alpha}^{A(p)}\}$, thus facilitating discussion of the relative importance of the relaxation promoting effects reflected by the off-diagonal components. It also partitions the *in situ* AIM FF index into its p-mode contributions, analogs of the $w_{i,\alpha}^{(p)}$ components of Eq. (198), which sum up to zero:

$$\sum_i \sum_\alpha \omega_{i,\alpha}^{CT(p)} \equiv \sum_\alpha \omega_\alpha^{CT} = \sum_i f_i^{CT} = \sum_a^A f_a^{CT} + \sum_b^A f_b^{CT} = 0 \qquad (205)$$

The partial sum,

$$\omega_\alpha^{CT(p)} = \frac{\partial p_\alpha}{\partial N_{CT}} \frac{\partial N}{\partial p_\alpha} \equiv F_\alpha^{CT(p)} \varphi_\alpha^{(p)} \qquad (206)$$

represents the CT analog of the $w_\alpha^{(p)}$ quantity of Eqs. (199) and (202). Like the partitioning of Eq. (202), it can be divided into the corresponding reactant contributions:

$$\omega_\alpha^{CT(p)} = \sum_{X=A,B} \frac{\partial p_\alpha}{\partial N_{CT}} \frac{\partial N_X}{\partial p_\alpha} = \sum_{X=A,B} F_\alpha^{CT(p)} \varphi_\alpha^{X(p)} \equiv \sum_{X=A,B} \Omega_{X,\alpha}^{CT(p)} \qquad (207)$$

Therefore, the overall mode quantity $\omega_\alpha^{CT(p)}$ represents the net change in the global number of electrons in M per unit CT, as a result of electron transfer between the reactants via the p-channel α; its reactant-resolved components monitor the corresponding changes in the overall numbers of electrons in A and B.

5 Two-Reactant Populational Reference Frames

5.1 Rigid and Relaxed Internal Hardness Decoupling Modes

Let us consider the *intra-reactant decoupling*, in which we rotate the AIM reference frame of M, $\mathbf{\partial}_M$, into the PNM of the mutually closed reactants, called the *internally decoupled modes* (IDM). If the relaxational influence of one reactant upon another is ignored, one readily obtains the rigid IDM by diagonalizing the reactant blocks of η (see Eq. (137)):

$$U_X^\dagger \eta^{X,X} U_X = h_X \quad \text{(diagonal)} \quad U_X^\dagger U_X = I_X, \quad X = A, B \qquad (208)$$

The columns of the resulting overall intra-reactant decoupling-transformation matrix (see Eq. (137)) define the rigid IDM. The orthogonal axes $\mathbf{U}^{int} = \mathbf{\partial}_M U_{int}$ represent the new basis vectors in this representation. The transformed hardness matrix assumes the partly decoupled form:

$$\eta_{int} \equiv U_{int}^\dagger \eta U_{int} = \begin{pmatrix} h_A & \eta_{int}^{A,B} \\ \eta_{int}^{B,A} & h_B \end{pmatrix} \qquad (209)$$

with $\eta_{int}^{A,B} = U_A^\dagger \eta^{A,B} U_B = (\eta_{int}^{B,A})^\dagger$. Obviously, the zero off-diagonal blocks in U_{int} reflect the rigid electron distribution in the environment of the primary displaced reactant assumed in IDM.

The corresponding *relaxed* IDM are defined by the eigenvalue problem of Eq. (134), which defines the overall relaxed transformation matrix $\tilde{\mathbf{U}}_{int}^{rel}$. This transformation rotates the AIM populational vectors into the relaxed IDM basis vectors, $\vec{U}_{rel}^{int} = \boldsymbol{\theta}_M \tilde{\mathbf{U}}_{int}^{rel}$.

In Fig. 7, the relaxed IDM of toluene in the chemisorption complex $M = (\text{toluene} | V_2O_5)$, involving the two-pyramid cluster of the vanadium pentoxide, are shown together with the AIM relaxed FF diagram and those representing its resolution into the CT- and P-components. These relaxed charge-distribution patterns should be compared with the corresponding separated reactant analogs (see Sect. 2.2.1). Such a comparison reveals a substantial

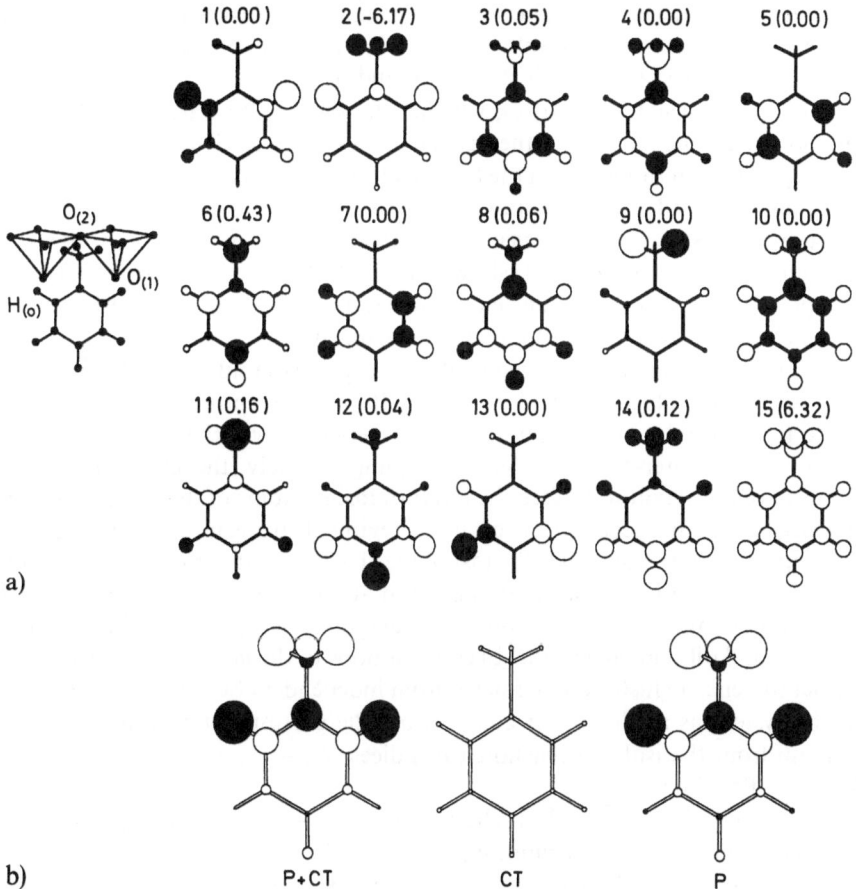

Fig. 7. Diagrams of the relaxed IDM (**Part a**) and the relaxed AIM FF indices (**Part b**) of toluene in the surface complex M with the bipyramidal cluster of the vanadium oxide surface. The modes are arranged in accordance with increasing hardnesses h_α^M; the hardness tensor reflects the isolated reactant AIM charges from the MNDO and scaled-INDO calculations on toluene and cluster, respectively. Numbers in parentheses report the w_α^M values. The three diagrams in **Part b** display the AIM FF distribution of toluene in M, calculated from the toluene block of $\boldsymbol{\eta}^{rel}$, and its resolution into the CT and P components, respectively

toluene-promotion effect due to the interaction with the oxide cluster. For example, three new positions exhibiting negative f_i^M indices are observed when toluene interacts with the surface, now also including the ring carbon bonded to the methyl group and the ortho-hydrogens, $H_{(o)}$, which strongly interact with the vanadyl oxygens $O_{(1)}$ at the surface.

The effect of relaxational promotion of the vanadium oxide cluster, due to the interaction with toluene, is illustrated in Fig. 8, where the unrelaxed PNM and FF data are compared with the corresponding IDM and the relaxed FF results. It follows from the figure that the hardness spectrum and the mode topologies are only slightly changed. However, the AIM FF diagram exhibits a dramatic change as a result of interaction, since the terminal $O_{(1)}$ oxygens have acquired the negative values in M. Again, there is less of the CT and more of the P components in the resultant (P + CT) FF indices of the relaxed reactant (compare the respective Panels in parts b and d).

The question naturally arises, whether the relaxed FF patterns of Figs. 7b and 8d match one another in the net toluene → cluster CT ($dN_{toluene} < 0$ and $dN_{cluster} > 0$), determined from the SCF MO calculations on M as a whole [44]. Since the reactant FF plots are normalized to 1, one has to change the phases of the toluene displacements (basic reactant) of Fig. 7b in the composite FF diagrams of the reactants in M, shown in Fig. 9, in order to obtain a common interpretation of shaded and unshaded circles; in such a diagram the FF pattern of the cluster (acidic reactant) remains unchanged.

Let us first examine the perpendicular adsorption of toluene on $O_{(2)}$ from the $O_{(1)}$ side of the cluster (part A of Fig. 9). It can be seen in Panels a–c of the figure that the relaxed FF patterns match perfectly over the whole region of a strong local inter-reactant charge coupling. Namely, the electron gaining atoms of one reactant in M face the electron losing atoms of the other reactant, and *vice versa*. Since local CT effects in the region of strong interactions are seen to be opposite to the overall CT, most of the charge donated by toluene (mainly from the $C_{(o)}$ positions) goes into vanadium atoms and the singly coordinated pyramid-base oxygens. The composite FF diagrams of Fig. 9 show the diagonal (Panel a) and off-diagonal (Panel b) contributions to the *in situ* FF (Panel c), for the net toluene → cluster CT predicted from independent SCF MO calculations on the system as a whole. For comparison, the corresponding composite FF diagram from the isolated reactant FF indices, $f_A - f_B$, (Figs. 3a and 8b), is shown in Panel d.

More specifically, Panel a shows the $f^{A,A} - f^{B,B} \equiv (\partial N_{cluster}/\partial N_{cluster})$ $- (\partial N_{toluene}/\partial N_{toluene})$ distribution, the $f^{A,B} - f^{B,A} \equiv (\partial N_{toluene}/\partial N_{cluster})$ $- (\partial N_{cluster}/\partial N_{toluene})$ displacements are reported in panel b, and their sum, $f_A^{CT} + f_B^{CT} \equiv (f^{A,A} - f^{B,A}) - (f^{B,B} - f^{A,B}) = (\partial N_{cluster}/\partial N_{CT}) + (\partial N_{toluene}/\partial N_{CT})$, is given in Panel c.

One observes immediately that the first two panels are practically indistinguishable. Thus, the diagonal and off-diagonal components of the relaxed *in situ* CT FF indices act in phase, giving the same perfect matching between reactants and enhancing one another in the overall indices of Panel c. Therefore, at least in

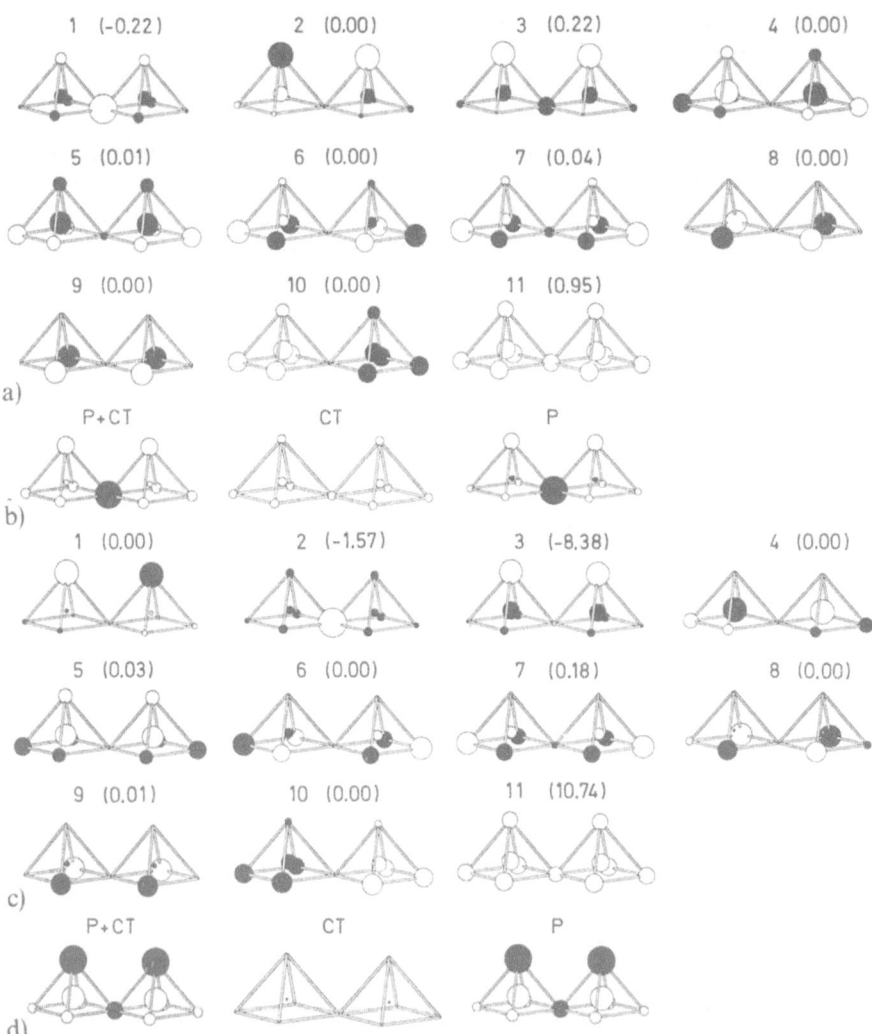

Fig. 8. Diagrams of the PNM (**Part a**) and the relaxed IDM (**Part c**) of the bipyramidal cluster of the vanadium oxide surface in complex M of Fig. 7a. The corresponding AIM FF plots are shown in **Parts b** and **d**, together with their resolutions into P and CT components. Modes are ordered in accordance with increasing $h_\alpha(h_\alpha^M)$ values (a.u.) reported at the bottom of the mode contour. Numbers in parentheses are the corresponding $w_\alpha(w_\alpha^M)$ values

this specific case, each type of CT FF indices may serve as a sensitive, adequate, two-reactant criterion for diagnosing the CT-related charge reorganization in M.

The isolated-reactant FF data give the composite FF diagram shown in Panel d, which also matches quite well in the interaction region. However, there is only a minor differentiation between peripheral atoms and those in the interaction region, with most of the toluene (cluster) atoms losing (gaining) electrons as a result of the toluene → cluster CT. The only local CT effect, which reverses this global picture, is the predicted $O_{(2)} \rightarrow C(H_3)$ charge shift. The relaxed FF data give rise to substantial differentiation among the ring atoms and in the $C_{(o)}-H_{(o)}$ bonds; none of these "activation" effects can be predicted from the isolated reactant data. Another difference between the interacting and separated reactant diagrams is that the former predicts that the largest charge reorganization should occur in the vicinity of strong inter-reactant charge

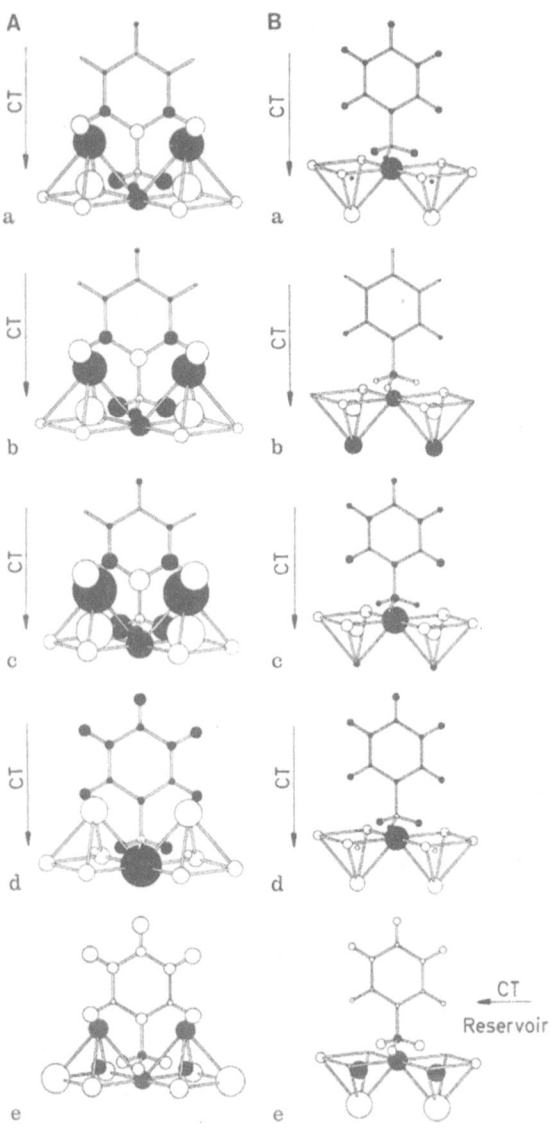

coupling, with only minor adjustments being observed on peripheral atoms, while the latter involves all the constituent atoms equally, most of them acting "in phase" with the assumed action of each reactant. The interacting reactant diagrams resemble more the minimum energy charge displacements, which exhibit the same general trend of a fast decay of charge shifts away from the primary populational perturbation, with an appreciable short-range polarization marked by a large number of nodes.

Panel e shows the AIM FF diagram for $M = (A|B)$ in contact with an external reservoir ($N \neq$ const.). As in the preceding diagrams, the AIM charges before the $B \rightarrow A$ CT have been used, in order to facilitate a comparison with the in situ ($N =$ const.) FF contours. Reference to Panels c, e of the figure indicates that the AIM FF pattern of M as a whole for $dN > 0$, resembles the in situ distribution in the region of strong interaction between $H_{(o)}$ and $O_{(1)}$ atoms, while the methyl group charge displacements exhibit the opposite phases in both cases. Notice, that for $dN < 0$ (i.e., for the opposite phases in Panel e) the open-system diagram closely resembles the FF plot of Panel d. This indicates, that an outflow of electrons from M (its chemical oxidation) enhances the toluene $\rightarrow$ cluster coordination, i.e., strengthens the overall effect of the toluene $-$ cluster chemisorption bond. Thus, the opposite influence, weakening of this bond, is to be expected for an inflow of electrons to M (its chemical reduction); this should facilitate toluene desorption.

Consider now the second (B) group of panels in Fig. 9, which corresponds to the perpendicular coordination of toluene to $O_{(2)}$ from the side of pyramid bases of the vanadium oxide cluster. In this case the charge coupling between reactants is much weaker, so that the FF pattern is dominated by the AIM charge displacements on the cluster (softer reactant). An inspection of Panels a and b reveals that some AIM displacements of the diagonal and off-diagonal contributions, e.g., on the methyl hydrogens and $O_{(1)}$ atoms, exhibit opposite phases, thus cancelling each other to some degree in the overall in situ diagram of Panel c. This resultant CT FF plot mainly exhibits transfer of electrons from the whole [toluene–$O_{(2)}$] unit to vanadium atoms and the remaining pyr-

Fig. 9. The AIM FF diagrams for the two perpendicular coordinations of toluene to the bridging oxygen $O_{(2)}$ from the $O_{(1)}$ side [**Panels A (a–e)**] and from the side of pyramid bases [**Panels B (a–e)**] of the bipyramidal cluster of vanadium oxide. For each case the matchings are shown between the basic (toluene) and acidic (cluster) diagonal relaxed FF indices (**Panels a**), the off-diagonal relaxed FF indices (**Panels b**), and the resultant in situ CT FF indices (**Panels c**, the sum of diagrams a and b) in the surface complex M of Fig. 7a (see Figs. 7b and 8d). In **Panels a** and **b** the phases of the atomic displacements on toluene (basic reactant) have been changed in order to achieve a common interpretation of populational displacements of both reactants [negative (shaded circles) and positive (open circles)] during the toluene $\rightarrow$ cluster CT. One notices that the diagonal and off-diagonal contributions to the CT FF act in phase. It follows from all of these diagrams that the local internuclear CT effects, $O_{(1)} \rightarrow H_{(o)}$ and $O_{(2)} \rightarrow C(H_3)$ are in the opposite direction relative to the global inter-reactant CT. A different picture emerges from the corresponding matching of the isolated reactant FF indices, $f_A - f_B$, shown in **Panels d**, where only the local $O_{(2)} \rightarrow C(H_3)$ charge shift reverses the overall CT between reactants. In **Panels e** the AIM FF diagram for M as a whole ($N \neq$ const.) is shown for comparison with the corresponding ($N =$ const.) diagrams

amid-base oxygens. Since the amount of charge removed from the bridging oxygen is large, one detects favourable conditions for the overall toluene $\rightarrow$ cluster coordination. There is a marked similarity between the *in situ* FF pattern of Panel c and that resulting from the isolated FF data of the reactants (Panel d); the only difference is observed in the $C(H_3)$ fragment and in the signs of the FF indices of the vanadyl oxygens. Generally speaking, the composite FF diagram of the isolated reactants resembles the diagonal, relaxed FF distribution of Panel a. Again, when one examines the FF pattern of the open adsorption complex shown in Panel e, one detects that many local CT trends are reversed relative to the isoelectronic diagram c. For $dN > 0$ (phases as shown in the figure) this charge reorganization reverses the CT effects of Panel c, thus weakening the adsorption bond and facilitating the associative desorption of toluene. Accordingly, for $dN < 0$ (opposite phases to those shown in figure) the original charge displacements and the *adsorption bond* are predicted to be strengthened.

Thus, relaxational promotion does indeed facilitate the intuitively expected charge reorganization accompanying the toluene $\rightarrow$ cluster CT, when the hypothetical barrier for it is removed. A comparison between Figs. (3a) and (7b) and Panels b and d of Fig. 8 shows that the very structure of the FF distribution has changed as a result of the other reactant's presence. Namely, in the isolated reactants, the CT & P components exhibit atomic populational shifts of comparable magnitude while the FF indices of the interacting reactants are strongly P-dominated.

Reference to Figs. 2 and 7a indicates that the IDM spectrum and FF indices exhibit dramatic changes in comparison with the corresponding PNM data of isolated toluene. The interaction with the V_2O_5 cluster has concentrated the CT-reactivity information in two modes (compare the w_α values in parentheses of Fig. 7a) : $\alpha = 2$ IDM, which dominates the CT-induced polarization (see the third diagram of Fig. 7b), and the nonselective, hardest mode $\alpha = 15$, which accounts for the external CT to the surface cluster (see the second diagram in Fig. 7b). Thus, a toluene–cluster interaction preselects the $\alpha = 2$ IDM (resembling the $\alpha = 12$ PNM of the isolated toluene) as the most important of the selective reactivity channels which differentiate between local sites in the adsorbed molecule. The topology of this crucial mode reflects a tendency of the physisorbed toluene to donate electrons to the cluster vanadium atoms through the methyl group hydrogens and to accept electrons (from the terminal $O_{(1)}$ oxygens of the cluster) through the ortho-hydrogens. A similar domination of the relaxed FF pattern of the vanadium oxide cluster is seen in Fig. 8c, where two modes, $\alpha = 3$ and 11, explain the main features of the first diagram in Panel d of the figure. Here again the role of mode $\alpha = 3$ has been strengthened as a result of the interaction, in comparison with the isolated cluster.

These illustrative results have profound implications for the theory of chemical reactivity and for catalysis in particular. They indicate that for strongly interacting, large reactants, the isolated-reactant charge-sensitivity data should be supplemented by the response criteria of the interacting species. One has to resort to this truly two-reactant information, which *preserves the memory of the actual interaction in the reactive system* as do the above IDM and their charge sensitivity characteristics. The presence of the other reactant filters out the most important reaction channels, which are not "recognized" by the charge-sensitivity criteria of the isolated species. The relaxational promotion of reactants, for their current separation and orientation in M, has been shown to generate an almost perfect matching of their relaxed FF data.

As we have already indicated, both the f and f^{CT} AIM FF indices can be partitioned in terms of contributions from collective charge-displacement modes. The corresponding rigid IDM partitionings of the AIM FF indices for the illustrative case of the reactive system M = toluene (B) – [V$_2$O$_5$-cluster] (A) are shown in Figs. 10 and 11.

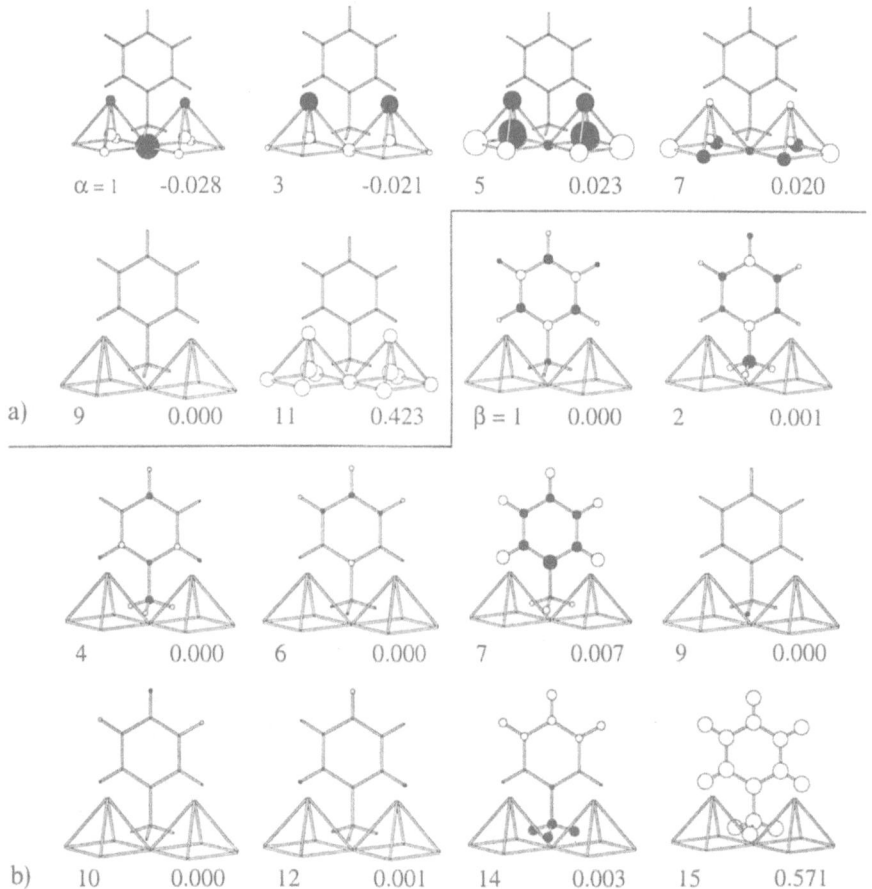

Fig. 10. The rigid IDM resolution (see Figs. 2 and 8a) of the AIM FF diagram of Fig. 9 A–e: **Part a** shows the $\{w_{a,\alpha}, (a, \alpha) \in A\}$ distribution contours, while **Part b** presents the corresponding $\{w_{b,\beta}, (b, \beta) \in B\}$ data. Only external CT-active modes are shown, for which $w_\alpha(w_\beta)$ (second number below the diagram) does not vanish identically (by symmetry)

A comparison between Figs. 9A–e and 10 shows that the resolution of f in the rigid IDM framework indeed leads to rather strong participation of relatively few PNM of the reactants. Namely, the cluster interaction part of f is seen to be already represented satisfactorily by the $\{w_{a,5}\}$ distribution in Fig. 10a, with the modes $\alpha = 1$ and 3 also strongly influencing the resultant pattern of Fig. 9A–e in this region of strong charge coupling with toluene. It should be noted that the remaining part of the cluster is also affected by modes $\alpha = 7$ and 11. The toluene part of the FF diagram of the whole system generally exhibits a similar degree of concentration of the IDM components. The region of strong interaction with vanadium oxide is seen to originate mainly from the $\{w_{b,\beta}\}$ contributions for $\beta = 2, 7, 14,$ and 15. In this case, however, the resultant pattern of Fig. 9A–e cannot be related to a single, crucial mode. As a rule, the dominant modes of the vanadium oxide cluster are relatively soft; the opposite is true for toluene, in which relatively hard modes determine the overall FF distribution.

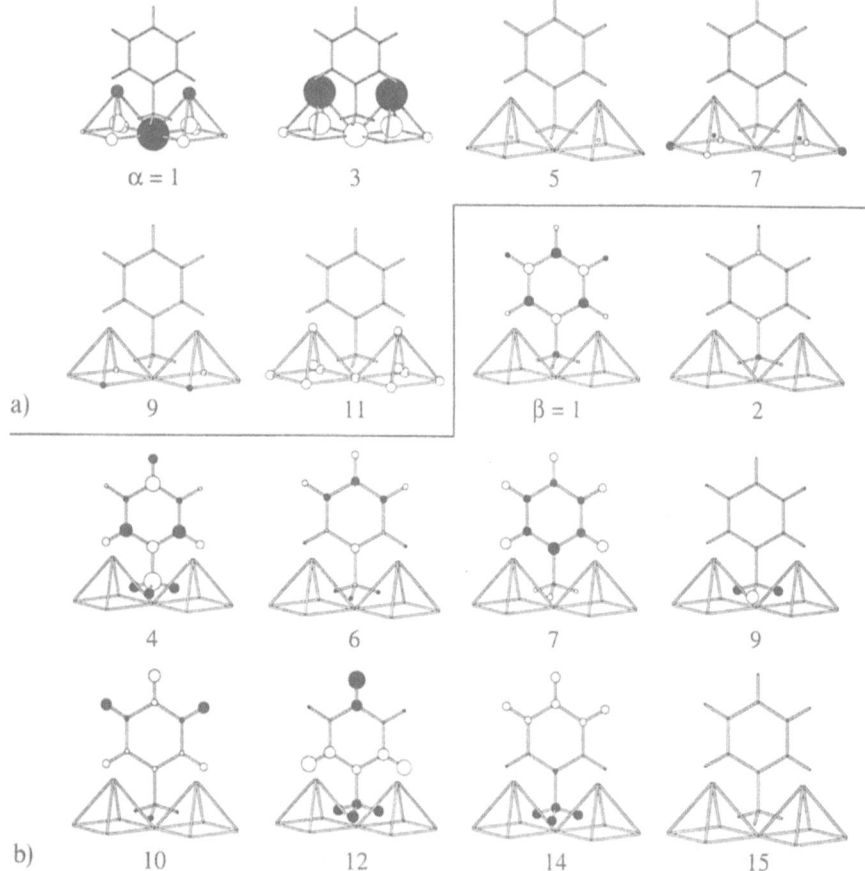

Fig. 11. The rigid IDM resolution, $\{\omega_{a,\alpha}^{CT}\}$- **Part a** and $\{\omega_{b,\beta}^{CT}\}$- **Part b**, of the AIM FF diagram of Fig. 9A–c

It follows from Fig. 11 that the rigid IDM resolution of f^{CT} (Fig. 9A–c) is more concentrated than that of Fig. 10. The cluster part of the toluene → cluster CT indices has large components $\omega_{a,\alpha}^{CT}$ only for the softest CT-active modes $\alpha = 1$ and 3. The toluene f^{CT} part in the strong interaction region is also relatively condensed, being dominated by the $\beta = 4$ and 12 modes, which alone explain the main qualitative features of Fig. 9a–c.

One concludes from this analysis that the rigid IDM provide an adequate reference frame for describing both external and internal CT processes in a reasonably compact form. However, carrying no information whatever about the reactant interaction, they are – by definition – one-reactant concepts, and therefore may not constitute the optimum collective charge displacement coordinates for reactive systems. In Sections 5.2 and 5.3 we extend this search into alternative modes of charge reorganization in M = A – – – B, the shapes of which are influenced by the actual interaction between A and B, similarly to the relaxed IDM.

In Figs. 12 and 13 we have examined the effect of relaxational contributions to the diagonal blocks of reactant hardness, due to the interaction with the other reactant. They present the relaxed IDM (see Figs. 7a and 8c) resolution of f and f^{CT}, respectively, in the same reactive toluene-[V_2O_5-cluster] system. A comparison with the previous, rigid IDM resolution of Figs. 10 and 11, respectively, indicates that the IDM components of f are only slightly modified by the mutual relaxational influence of reactants. f is dominated by the IDM exhibiting a similar topology and approximately the same $w_{\alpha(\beta)}$ values in both the rigid and relaxed cases. It is seen that the inclusion of relaxational effects still gives rise to dominance of both the toluene and cluster parts by the non-selective modes $\alpha = 11$ and $\beta = 15$ (Fig. 12), which together account for practically the total electron transfer to/from the system, whereas similarly selective modes determine the internal polarization within each reactant.

It follows from a comparison between Figs. 11 and 13 that the relaxational influence leads to a more compact resolution of the f^{CT} (toluene → cluster), particularly on toluene, where the accompanying charge redistribution is now well described by a *single* mode $\beta = 2$. Again, the two modes, $\alpha = 2, 3$ dominate the cluster part, with the former accounting for the negative $f_{O_{(2)}}^{CT}$ and the latter reproducing the negative $f_{O_{(1)}}^{CT}$ and positive f_V^{CT} values. This result again demonstrates that the interaction-dependent reference frame is better suited for describing the internal CT between reactants in M than the corresponding isolated-reactant PNM coordinate system.

5.2 Externally Decoupled Modes from the Maximum Overlap Criterion

Another, complementary set of partially decoupled populational modes is defined within the external hardness decoupling scheme [9, 45]. The externally decoupled mode (EDM) transformation, $\tilde{U}^{ext} = \tilde{e}_M U_{ext}$, transforms the AIM hardness matrix η of the reactive system $M^+ = (A^+ | B^+)$ into the block-

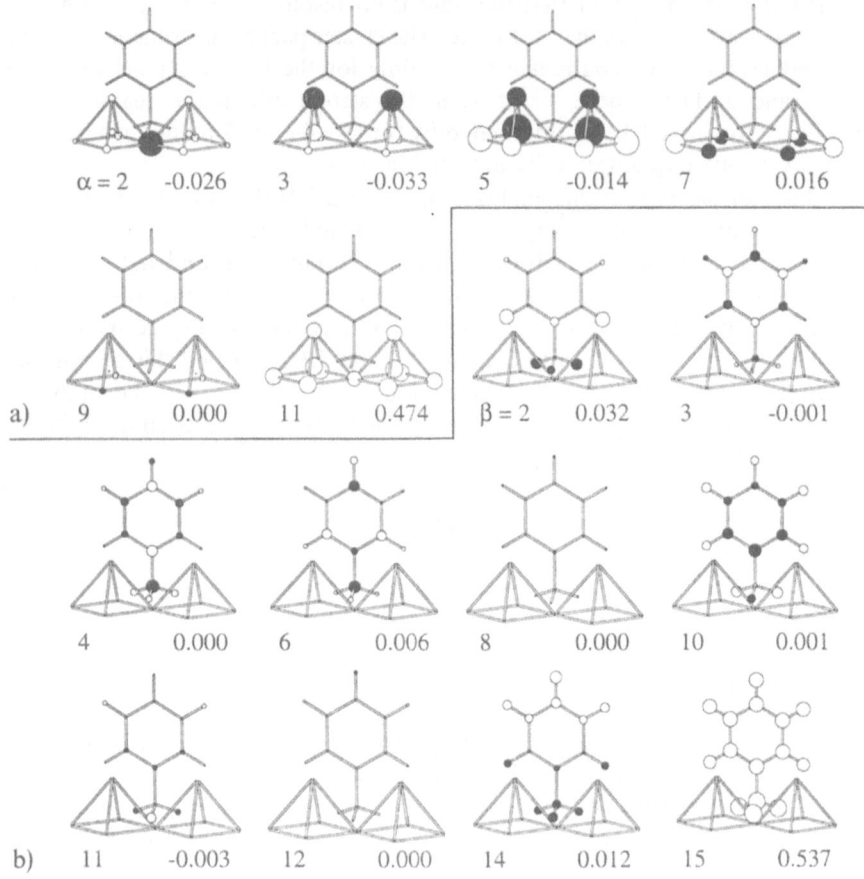

Fig. 12. The relaxed IDM resolution of f (Fig. 9 A–e); see also caption to Fig. 10.

diagonal form:

$$\boldsymbol{\eta}_{\text{ext}} \equiv \mathbf{U}^{\dagger}_{\text{ext}} \boldsymbol{\eta} \mathbf{U}_{\text{ext}} = \begin{pmatrix} \boldsymbol{\eta}^{A,A}_{\text{ext}} & \mathbf{O}^{A,B} \\ \mathbf{O}^{B,A} & \boldsymbol{\eta}^{B,B}_{\text{ext}} \end{pmatrix} \qquad (210)$$

in which the off-diagonal, interaction block vanishes, $\boldsymbol{\eta}^{A,B}_{\text{ext}} = \mathbf{O}^{A,B}$. Such transformation is not unique. A particular case of $\mathbf{U}_{\text{ext}} = \mathbf{U}\mathbf{U}^{\dagger}_{\text{int}}$, where $\mathbf{U}$ is the total decoupling transformation of Eq. (60), has been tested numerically for the toluene–$[V_2O_5]$ and ethylene–butadiene reactive systems [45]. However, the modes generated by this transformation were delocalized throughout M, with no direct reference to a particular reactant, which – in effect – prevented them from being used to described the mode-to-mode CT interaction between reactants in M^+.

Consider the following two-step algorithm for generating the *localized* (ℓ) EDM, $\bar{\mathbf{U}}^{\text{ext}}_{(\ell)} = (\bar{\mathbf{U}}^{\text{ext}}_A, \bar{\mathbf{U}}^{\text{ext}}_B)$, where $\bar{\mathbf{U}}^{\text{ext}}_X$ represents the EDM predominantly lo-

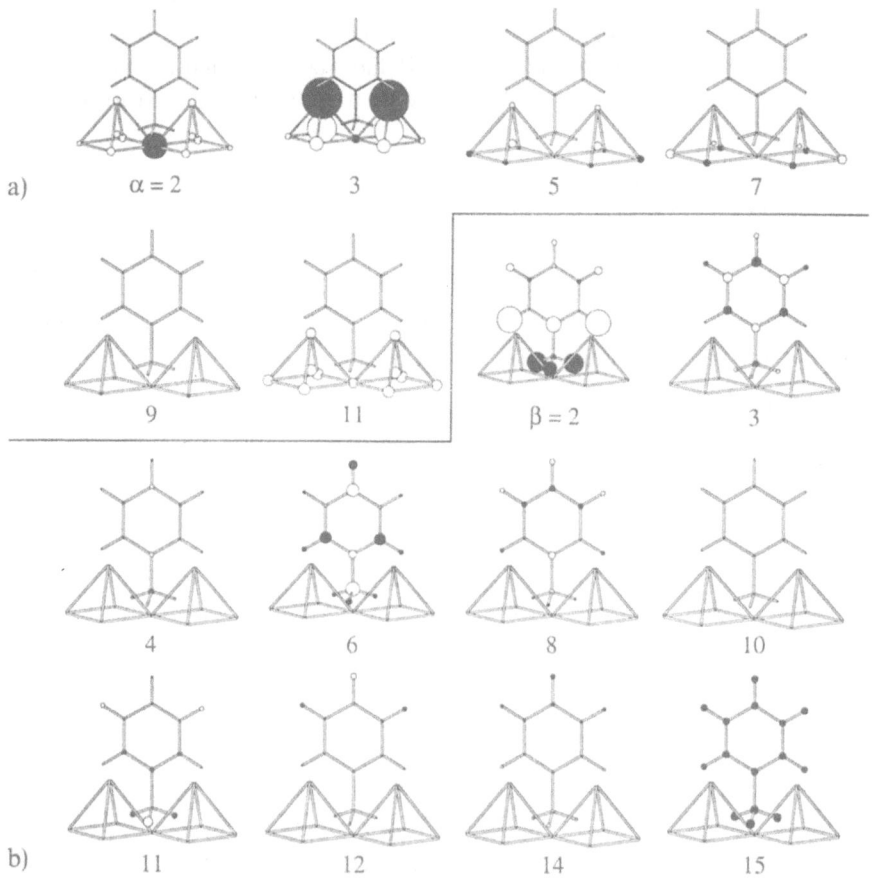

Fig. 13. The relaxed IDM resolution of f^{CT} (Fig. 9 A–c).

calized on $X = (A, B)$. It also involves the intermediate stage of the total decoupling of η;

$$\vec{U}^{ext}_{(\ell)} = (\vec{e}_M U) U' = \vec{U}_M U' \tag{211}$$

and the block-diagonal, unitary *localization transformation*,

$$U' = \begin{pmatrix} U'_{A,A} & O^{A,B} \\ O^{B,A} & U'_{B,B} \end{pmatrix} \tag{212}$$

which automatically guarantees the desired block structure of η_{ext}. U' is determined from the *maximum overlap criterion* (MOC) [46]

$$\mathrm{tr}(\vec{U}^{int})^{\dagger} \vec{U}^{ext}_{(\ell)} = \text{maximum} \tag{213}$$

We first divide the PNM set $\vec{U}_M$ into two subsets, $\vec{U}_M = (\vec{U}_A, \vec{U}_B)$, defined by the corresponding $m \times n$ and $m \times (m - n)$ rectangular matrices in

$\mathbf{U} = (\mathbf{U}^A | \mathbf{U}^B)$, using the criterion of maximum overall projection $P(\alpha \to X) = \Sigma_i^X |U_{i\alpha}|^2$ of the mode α onto the subspace $\boldsymbol{\vartheta}_X$ of reactant $X = (A, B)$. Namely, the columns of $\mathbf{U}^A$ and $\mathbf{U}^B$ have been selected to maximize the $P(\alpha \to A)$ and $P(\beta \to B)$, respectively. We then apply the reactant localization transformations:

$$\vec{U}_X^{ext} = \vec{U}_X \mathbf{U}_{X,X}^{\varsigma}, \quad X = (A, B) \tag{214}$$

which simultaneously maximize

$$\mathrm{tr}(\vec{U}_X^{int})^\dagger \, \vec{U}_X^{ext} = \mathrm{tr}\Lambda_X, \quad X = (A, B) \tag{215}$$

where $\vec{U}_X^{int} = \boldsymbol{\vartheta}_X \mathbf{U}_X$. As shown by Gołębiewski [46], this localization transformation is defined by the following expression:

$$\mathbf{U}_{X,X}^{\varsigma} = \boldsymbol{\Delta}_X^\dagger (\boldsymbol{\Delta}_X \boldsymbol{\Delta}_X^\dagger)^{-1/2}, \quad X = (A, B) \tag{216}$$

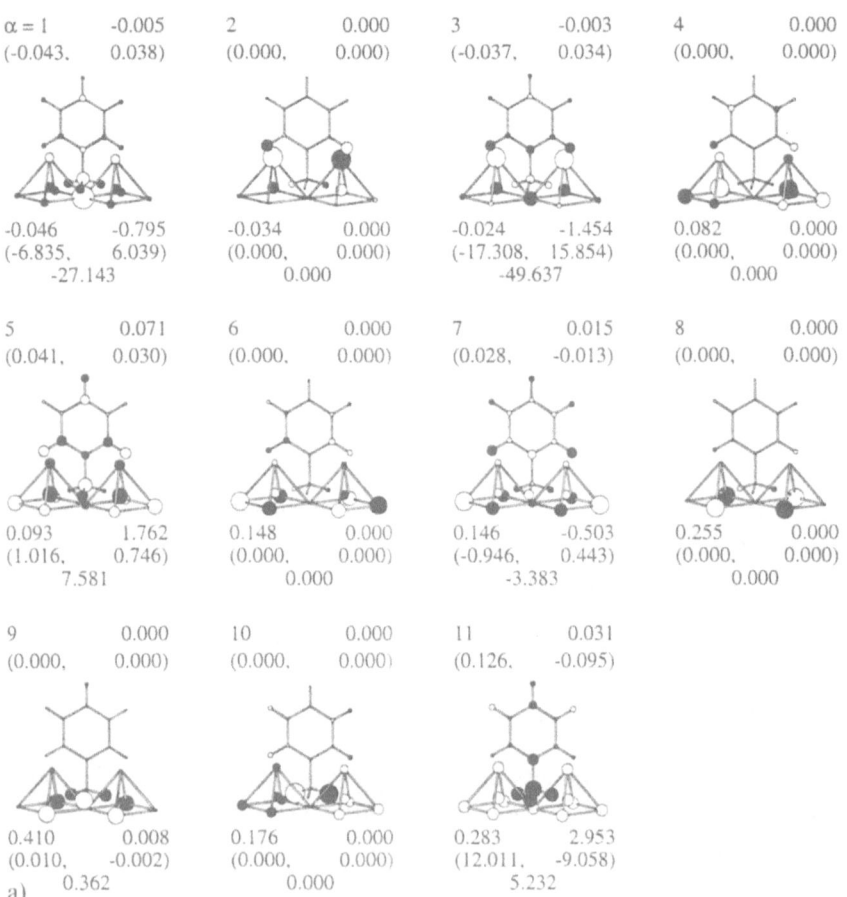

$\alpha = 1$	-0.005	2	0.000	3	-0.003	4	0.000
(-0.043,	0.038)	(0.000,	0.000)	(-0.037,	0.034)	(0.000,	0.000)

-0.046	-0.795	-0.034	0.000	-0.024	-1.454	0.082	0.000
(-6.835,	6.039)	(0.000,	0.000)	(-17.308,	15.854)	(0.000,	0.000)
	-27.143		0.000		-49.637		0.000

5	0.071	6	0.000	7	0.015	8	0.000
(0.041,	0.030)	(0.000,	0.000)	(0.028,	-0.013)	(0.000,	0.000)

0.093	1.762	0.148	0.000	0.146	-0.503	0.255	0.000
(1.016,	0.746)	(0.000,	0.000)	(-0.946,	0.443)	(0.000,	0.000)
	7.581		0.000		-3.383		0.000

9	0.000	10	0.000	11	0.031
(0.000,	0.000)	(0.000,	0.000)	(0.126,	-0.095)

0.410	0.008	0.176	0.000	0.283	2.953
(0.010,	-0.002)	(0.000,	0.000)	(12.011,	-9.058)
a)	0.362		0.000		5.232

Fig. 14. (Begining).

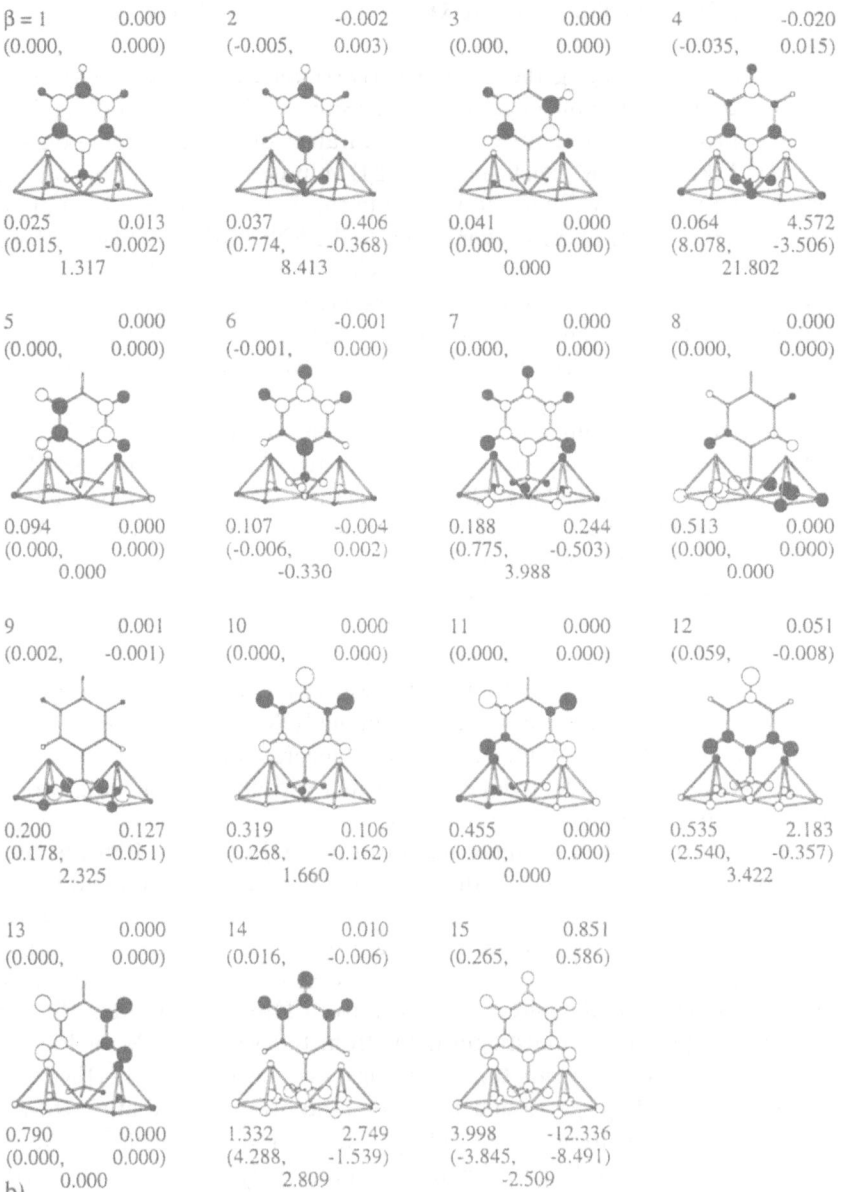

Fig. 14. (Continued). The localized EDM for the M = toluene–[V$_2$O$_5$] system, numbered in accordance with the isolated reactant IDM they resemble the most (see Figs. 2 and 8a). The usual phase convention $\varphi_\alpha^{(EDM)} \equiv \partial N/\partial p_\alpha > 0$ has been adopted. **Part a** groups the EDM associated with the (V$_2$O$_5$) cluster, X = A, and **Part b** corresponds to toluene, X = B. The numerical data above each diagram report the number of IDM giving the maximum projection on the EDM in question, $w_\alpha^{(EDM)}$, followed by its resolution into reactant contributions $w_\alpha^{A(EDM)}$ and $w_\alpha^{B(EDM)}$, in parentheses. The bottom numbers include: $(\eta_{ext}^{X,X})_{\alpha,\alpha}$, $\omega_\alpha^{CT(EDM)}$ (first row), $\Omega_{A,\alpha}^{CT(EDM)}$, $\Omega_{B,\alpha}^{CT(EDM)}$ (second row, in parentheses), and $F_\alpha^{CT(EDM)}$

$\Delta_X = (\bar{U}_X^{int})^\dagger \bar{U}_X = (U_X)^\dagger U^{X,X}$, where $U^{X,X}$ is the corresponding diagonal block of U.

This way of generating the EDM ($\rightarrow$ IDM) collective modes guarantees their one-to-one correspondence to the respective IDM of reactants and, as such, may provide an alternative, convenient framework for describing the CT processes. Like the IDM, such localized EDM preserve the memory of the reactant interaction in M^+ and should lead to substantial hardness decoupling. This expectation is due to their resemblance to the PNM (IDM) of reactants, for which the diagonal (reactant) blocks of the relevant hardness matrix are exactly diagonal. Thus, with no external hardness interaction between the A and B subsets of EDM, it comes as no surprise that these collective, delocalized charge-displacement modes also bear some similarity to the PNM of M as a whole.

In addition to a dominant component of one reactant, each EDM also exhibits the matching component of the other reactant. Therefore, the EDM provide a mode-mode matching between reactants; as practically independent, with only a small hardness coupling within each subset, they should be useful in an interpretation of the overall B $\rightarrow$ A CT providing the mode-to-mode perspective on the accompanying charge reorganization in the reactants.

Illustrative localized EDM ($\rightarrow$ IDM) for M = (toluene–$[V_2O_5]$) are shown in Fig. 14, with the sets (a) and (b) grouping the subsets $\{\alpha\}$ and $\{\beta\}$ associated with the $[V_2O_5]$ cluster (A) and toluene (B), respectively. A comparison between these CT-reactive channels and the corresponding IDM of reactants (Figs. 2 and 8a) shows that the MOC adopted does indeed generate the inter-set decoupled displacements, which can be uniquely associated with the corresponding IDM of the separated reactants.

It is also of interest to compare the localized EDM and the PNM (subset $\{\gamma\}$) of M as a whole (Fig. 15), and the corresponding diagrams representing the collective-mode partitioning of the AIM FF indices (Figs. 16 and 17, respectively). It follows from the $w_\alpha^{(p)}$ values reported in Figs. 14, 16, and 17, that, when the system acts as a whole in an external CT process, the mode contributions are strongly dominated by the hardest, non-selective EDM, $\beta = 15$ (Figs. 14b and 16b), which is practically indistinguishable from the $\gamma = 26$ PNM of Fig. 17. Next in this $w_\alpha^{(p)}$ hierarchy are the $\alpha = 5$ (Figs. 14a and 16a), $\beta = 12$ (Figs. 14b and 16b), and $\gamma = 20$ (Figs. 15 and 17) modes; important contributions to the FF pattern is also due to modes $\alpha = 11$, $\beta = 4$, and $\gamma = 1$. As explicitly shown in Figs. 16 and 17, the $w_\alpha^{(p)}$ parameters do not fully reflect the mode participation in shifting the AIM charges, since some polarizational modes ($w_\alpha^{(p)} \approx 0$) also exhibit large amplitudes in their AIM electron population displacements. When one is interested in channels for shifting the electrons between reactants, their relative importance is reflected by the $w_\alpha^{A(p)}$ and $w_\beta^{B(p)}$ quantities, shown in parentheses above each mode diagram.

It follows from a comparison between Figs. 14, 16 and Figs. 15, 17, respectively, that the localized EDM already strongly resemble the totally decoupled PNM. In most cases, a clear one-to-one correspondence can be established by

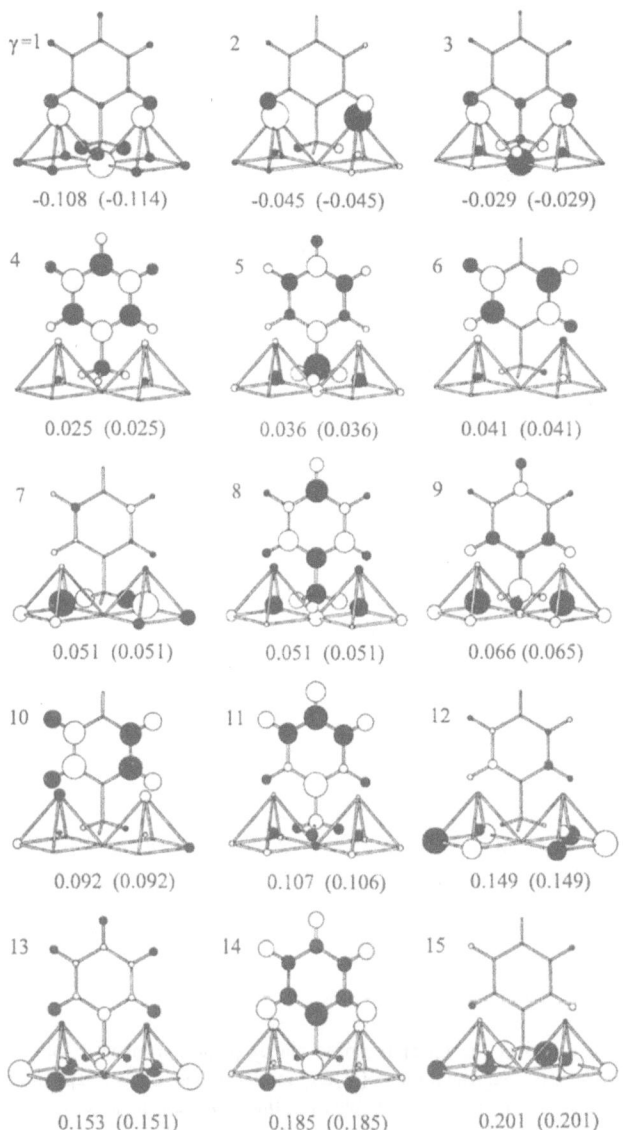

Fig. 15. (Begining).

inspection. Compare, e.g., the CT-active modes of Figs. 16 and 17. The $\alpha = 1, 3, 5, 7, 9$, and 11 EDM patterns of Fig. 16a may be related to those corresponding to $\gamma = 1, 3, 9, 13, 21$, and 20 PNM of Fig. 17, respectively. Similarly, the most important EDM patterns $\beta = 1, 2, 4, 6, 9, 12, 14$, and 15 of Fig. 16b are seen to resemble closely those observed for $\gamma = 4, 5, 8, 11, 16, 22, 25$, and 26, in Fig. 17.

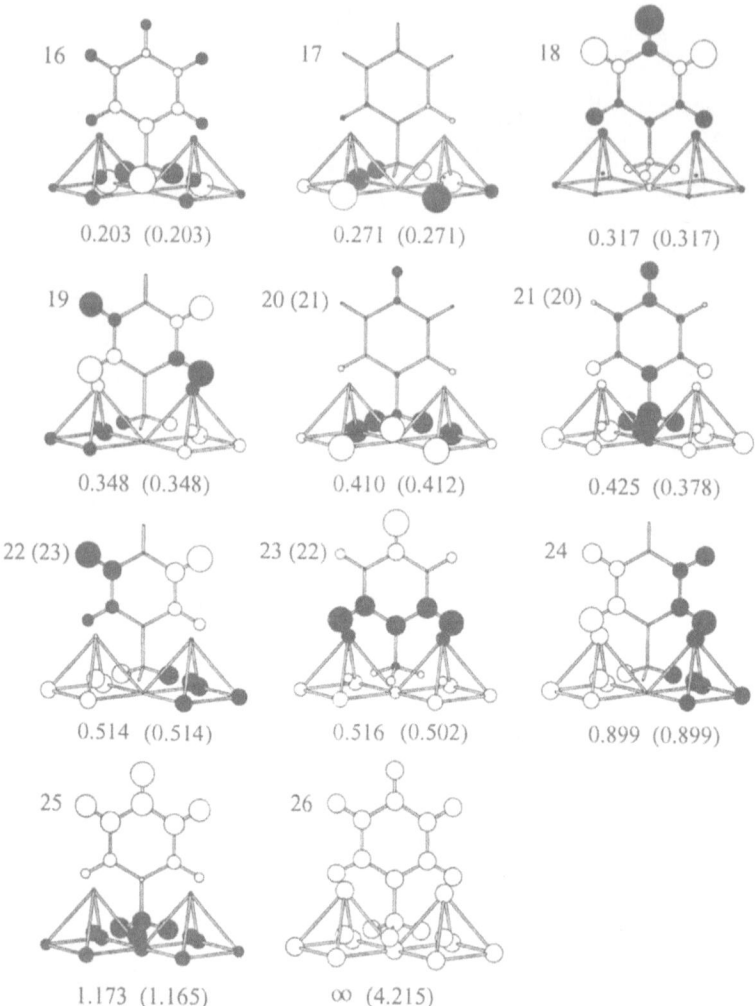

Fig. 15. (Continued). The eigenvectors of the linear response matrix β (or the internal softness matrix), the *internal normal modes* (INM) in the toluene–$[V_2O_5]$ model system; the corresponding PNM, i.e., eigenvectors of η are practically indistinguishable from these softness eigenvectors. The mode numbering is in accordance with the mode hardness eigenvalue (inverse of the corresponding eigenvalue of β). In cases where the ordering of these modes differs from that of PNM, the corresponding PNM number (in order of increasing hardness) is given in parentheses. The numbers reported at the bottom of each diagram are the inverse of the β-eigenvalue and the principal hardness of η. The nonselective, hardest (pure CT) mode exhibits an infinite β-hardness eigenvalue, and is thus excluded from the remaining internal polarizational modes

We therefore conclude that our algorithm for the IDM-*localized externally decoupled modes*, EDM ($\rightarrow$ IDM), does indeed generate reactive charge-displacement modes which can be easily correlated with both the IDM of reactants and the PNM of M as a whole. It follows from our illustrative results

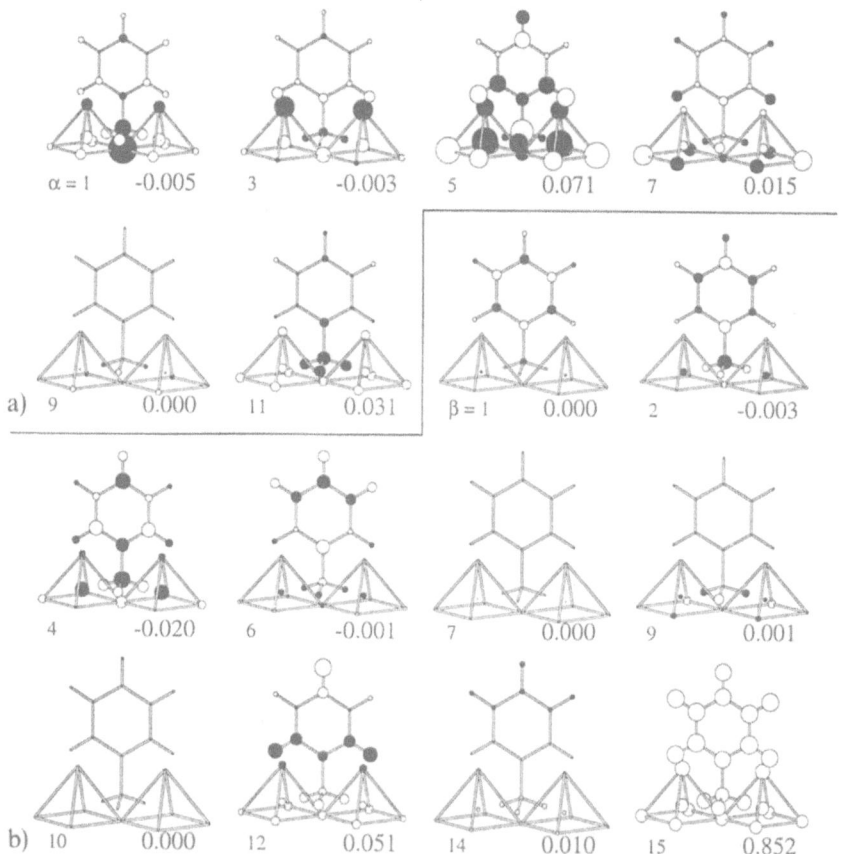

Fig. 16. The EDM resolution, $\{w_{i,\alpha}^{(EDM)}\}$, of the FF diagram of Fig. 9 A–e; only the CT-active ($w_{\alpha}^{(EDM)} \neq 0$, shown below the mode diagram) modes are reported. The mode numbers are the same as in the corresponding parts of Fig. 8

that the EDM ($\to$ IDM) set gives rise to a strong concentration of the AIM-resolved FF information in a few crucial modes; this mode-resolution closely follows that corresponding to the PNM description. The remarkable similarity between PNM and these localized EDM also indicates that there is little intra-set hardness coupling within the $\{\alpha\}$ and $\{\beta\}$ sets. This prediction can also be confirmed numerically.

Reference to the EDM hardnesses displayed in Fig. 14 (the first number under each diagram) shows that the three softest EDM, $\alpha = (1-3)$ are unstable $[\eta_{ext}^{A,A}]_{\alpha\alpha} < 0]$; they are dominated by the cluster charge displacements. In this mode-hardness ordering, the next four softest modes of the B-set, ($\beta = 1-4$), exhibit mostly the intra-toluene polarization. As in the PNM case, the modes representing the long-range polarization, with large groups of neighbouring atoms acting in the same phase, belong to the category of relatively hard

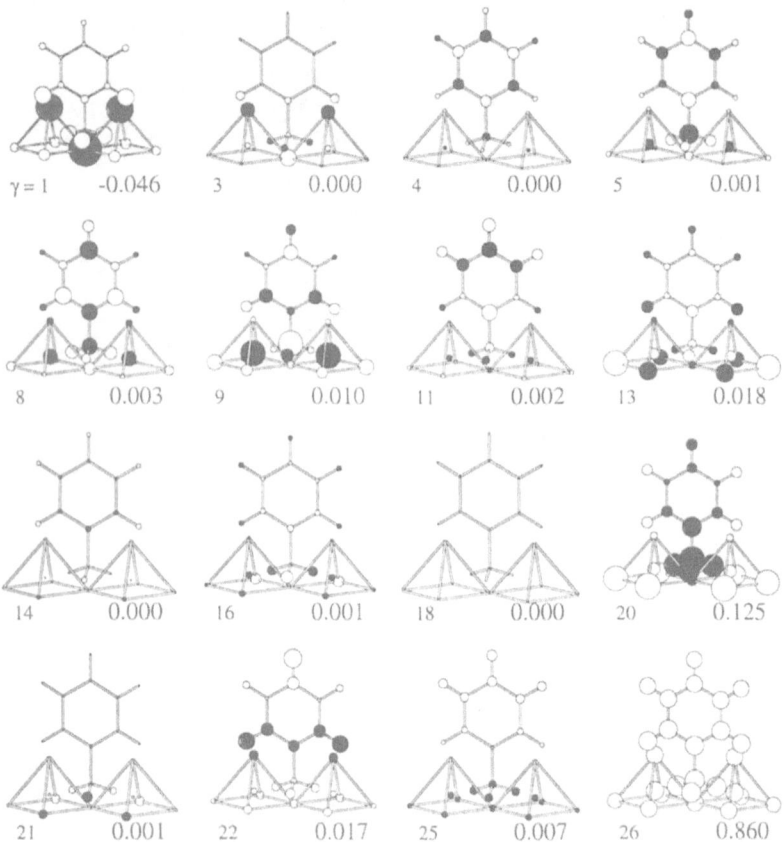

Fig. 17. The PNM resolution, $\{w_{i,\alpha}\}$, of the FF diagram of Fig. 9 A–e; only the CT-active ($w_\alpha \neq 0$) modes are shown. The mode numbers and the overall w_α index are reported at the bottom of each contour diagram

collective displacements. This is particularly so for modes representing CT between the reactants, e.g., $\beta = 12$, 14, or the external CT, $\beta = 15$.

The w_α^X (X = A, B) values are reported in parentheses above each mode diagram in Fig. 14. They show the fractions of dN channeled via mode α into both reactants. They also represent the importance of the mode in question in effecting CT between the reactants, since even small values of $|w_\alpha|$ may result from relatively large variations in the reactant's global number of electrons, as reflected by the w_α^X indices. This can be seen in modes $\alpha = 1, 3, 7, 11$ of Fig. 14a, which are the most important within the X = A subset for the CT involving an external reservoir (see also Fig. 16a).

One also observes that the most important external CT-active EDM in Fig. 14a (the largest $|w_\alpha|$ value), $\alpha = 5$, corresponds to a net increase in the number of electrons on both reactants when $dN > 0$ (inflow of electrons to M). The second mode of the figure in this $|w_\alpha|$ hierarchy, $\alpha = 11$, represents the pattern of electron redistribution between reactants which strengthens the toluene → cluster electron transfer when $dN > 0$.

Most of the change in dN via the non-selective mode, $\beta = 15$ in Fig. 14b, is on toluene, mainly because of the different numbers of constituent atoms in both reactants. Among the site-selective EDM in this figure, which are active in the external CT processes, the most important for the dN-induced charge redistribution (see Fig. 16b) are modes $\beta = 1, 2, 4, 6, 12$; the w_α^X indices of Fig. 14b show that modes $\beta = 1, 2, 6, 7, 10$ mainly redistribute electrons inside each reactant (small $|w_\alpha^X|$ values). For $dN > 0$, the inflowing electrons through the $\beta = 12$ mode are directed mainly to the cluster, while the $\beta = 4$ mode exhibits a net cluster → toluene CT.

Finally, let us examine the ω_α^{CT} quantities of the EDM and their reactant partitioning into $\Omega_{X,\alpha}^{CT}$, X = A, B. These parameters are displayed in Fig. 14, at the bottom of each diagram, in parentheses. These numerical values were generated for the assumed toluene (B) → cluster (A) electron transfer. Consider first the EDM that resembles the PNM of the cluster (Fig. 14a). Among the internal CT-active EDM, for which $\Omega_{X,\alpha}^{CT} \neq 0$, the $\alpha = 1, 3$, and 7 modes tend to decrease the global number of electrons ($\omega_\alpha^{CT} < 0$); the $\alpha = 5, 9$, and 11 EDM, for which $\omega_\alpha^{CT} > 0$, act in the opposite direction. For the $\alpha = 11$ channel, one finds $\Omega_{A,\alpha}^{CT} > 0$ and $\Omega_{B,\alpha}^{CT} < 0$, so that the mode CT is in the assumed direction; the remaining internal CT-active modes exhibit the opposite net CT effect: $\Omega_{A,\alpha}^{CT} < 0$ and $\Omega_{B,\alpha}^{CT} > 0$ ($\alpha = 1, 3, 7$).

In the second part of Fig. 14 one also finds that the majority of modes redistribute global electron numbers of reactants in the direction of the assumed CT ($\alpha = 2, 4, 7, 9, 10, 12, 14$). Again, it should be noted that the fully symmetrical background channel, $\alpha = 15$, represents a practically purely external CT, with both reactants loosing electrons in the assumed direction of CT.

Clearly, one can also define the EDM which resemble most closely the MEC/REC associated with the constituent atoms of both reactants, $\vec{r}_M = (\vec{r}_A, \vec{r}_B)$. Then, the corresponding localization transformation blocks, which maximize tr $\vec{r}_X^\dagger \vec{U}_X^{ext}$, X = (A, B), immediately follow from the last equation when Δ_X is appropriately redefined.

In Fig. 18 we have compared the external MEC in the model toluene-$[V_2O_5]$ system with the EDM localized to resemble the corresponding MEC: EDM (→ MEC). One should recall that the *non-orthogonal* MEC are all dominated by the softest, short-range polarization modes. Only the softest modes, in the *orthogonal* EDM set, localized mainly on the vanadium oxide cluster, exhibit such polarization, and thus indeed generally resemble the MEC set. One detects a degree of similar short-range polarization in the EDM of Part c. However, as seen in Part b, the EDM localized on the carbon MEC, are quite different from their MEC analogs, since – due to the EDM orthogonality constraints – the short-range PNM do not contribute to these relatively hard EDM.

These differences are reflected well by the mode hardness parameters. When a pair of (EDM, MEC) diagrams exhibits a similar topology, e.g., selected modes

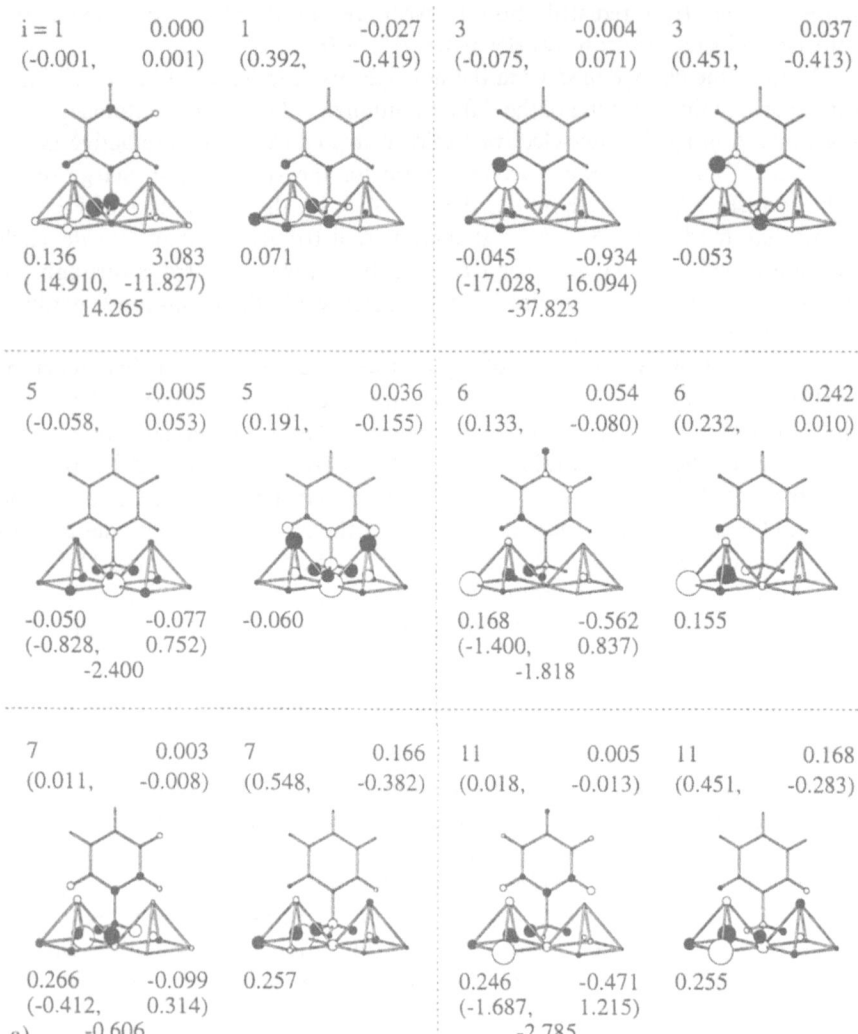

<table>
<tr><td>i = 1</td><td>0.000</td><td>1</td><td>-0.027</td><td>3</td><td>-0.004</td><td>3</td><td>0.037</td></tr>
<tr><td>(-0.001,</td><td>0.001)</td><td>(0.392,</td><td>-0.419)</td><td>(-0.075,</td><td>0.071)</td><td>(0.451,</td><td>-0.413)</td></tr>
</table>

0.136	3.083	0.071		-0.045	-0.934	-0.053
(14.910,	-11.827)			(-17.028,	16.094)	
	14.265				-37.823	

| 5 | -0.005 | 5 | 0.036 | 6 | 0.054 | 6 | 0.242 |
| (-0.058, | 0.053) | (0.191, | -0.155) | (0.133, | -0.080) | (0.232, | 0.010) |

-0.050	-0.077	-0.060		0.168	-0.562	0.155
(-0.828,	0.752)			(-1.400,	0.837)	
	-2.400				-1.818	

| 7 | 0.003 | 7 | 0.166 | 11 | 0.005 | 11 | 0.168 |
| (0.011, | -0.008) | (0.548, | -0.382) | (0.018, | -0.013) | (0.451, | -0.283) |

0.266	-0.099	0.257		0.246	-0.471	0.255
(-0.412,	0.314)			(-1.687,	1.215)	
a)	-0.606				-2.785	

Fig. 18. (Begining).

in Parts a and c, the hardnesses are of comparable magnitude. The opposite conclusion follows from Part b, where the hardnesses of the EDM are dramatically higher than those of their MEC analogs.

Another obvious conclusion that follows from Fig. 18 is that the EDM (→ MEC) are localized, as are the MEC modes that they resemble. This is particularly so in Part b of the figure, where modes $\alpha = 14$ and 15 represent the corresponding CH fragments of toluene. A reference to the w_α values, which reflect a partitioning of the external CT, dN, for the most important CT-active

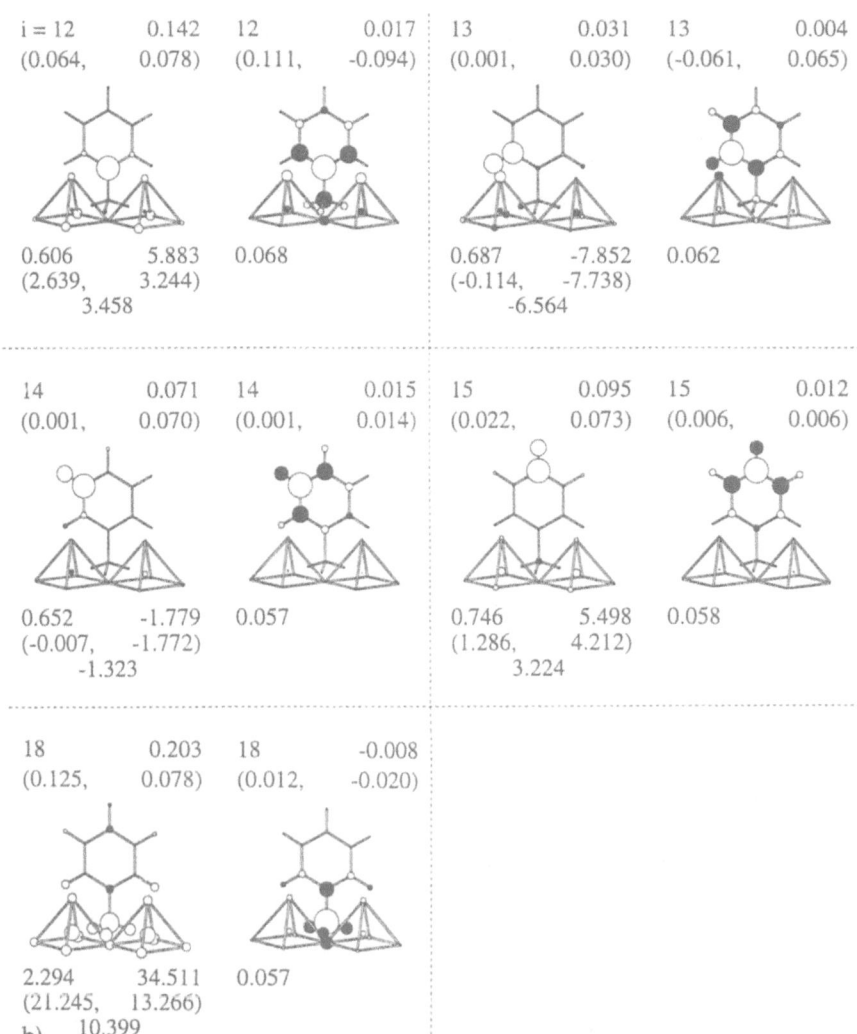

| i = 12 | 0.142 | 12 | 0.017 | 13 | 0.031 | 13 | 0.004 |
| (0.064, | 0.078) | (0.111, | -0.094) | (0.001, | 0.030) | (-0.061, | 0.065) |

0.606	5.883	0.068		0.687	-7.852	0.062
(2.639,	3.244)			(-0.114,	-7.738)	
	3.458				-6.564	

| 14 | 0.071 | 14 | 0.015 | 15 | 0.095 | 15 | 0.012 |
| (0.001, | 0.070) | (0.001, | 0.014) | (0.022, | 0.073) | (0.006, | 0.006) |

0.652	-1.779	0.057		0.746	5.498	0.058
(-0.007,	-1.772)			(1.286,	4.212)	
	-1.323				3.224	

| 18 | 0.203 | 18 | -0.008 |
| (0.125, | 0.078) | (0.012, | -0.020) |

2.294	34.511	0.057
(21.245,	13.266)	
b)	10.399	

Fig. 18. (Continued).

EDM ($\rightarrow$ IDM) (Fig. 16), PNM (Fig. 17) and the EDM ($\rightarrow$ MEC) (Fig. 18), clearly indicates that EDM ($\rightarrow$ IDM) give almost as compact description of the external CT phenomena as do PNM; in this respect the charge inflow (or outflow) from the system as a whole is seen to be dispersed into a relatively large subset of the EDM ($\rightarrow$ MEC), $\alpha = 18, 12, 15, 14, 25, 26, 6, 21,$ and 13, in decreasing order of $|w_\alpha|$. Therefore, the EDM ($\rightarrow$ IDM) appear to be a much more suitable collective populational coordinate system for describing the external CT than the EDM ($\rightarrow$ MEC) reference frame.

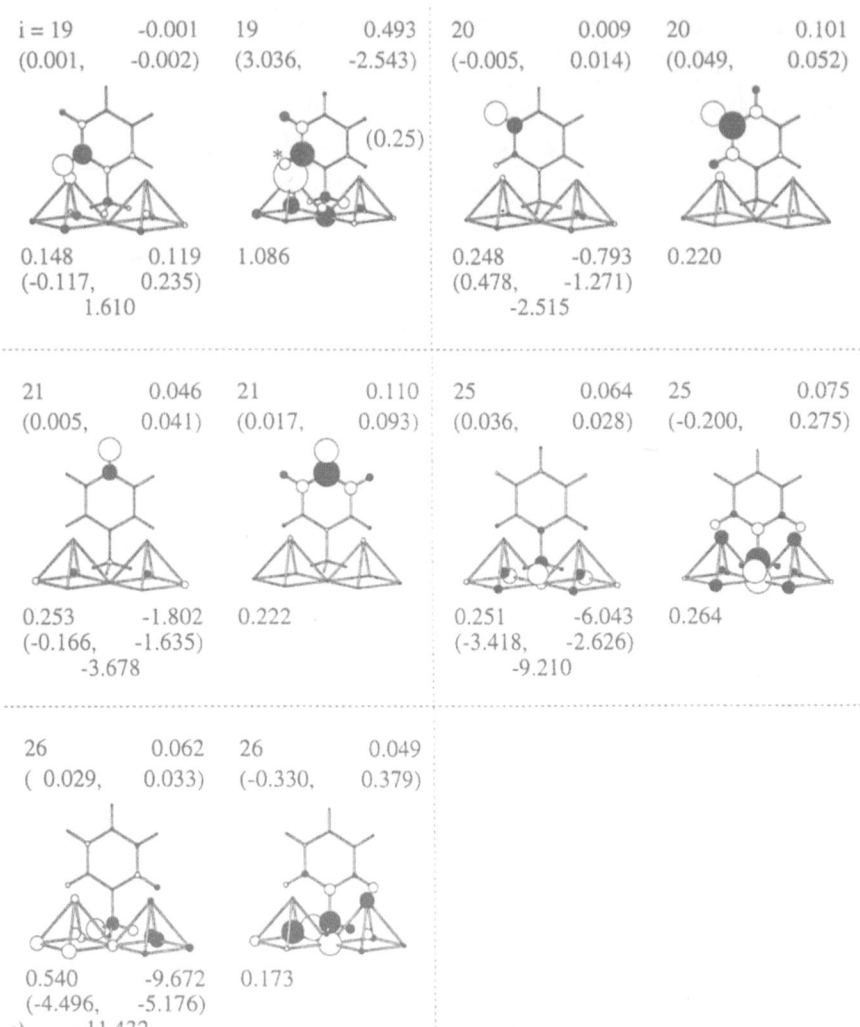

i = 19   -0.001   19   0.493   20   0.009   20   0.101
(0.001,   -0.002)   (3.036,   -2.543)   (-0.005,   0.014)   (0.049,   0.052)

(0.25)

0.148   0.119   1.086   0.248   -0.793   0.220
(-0.117,   0.235)   (0.478,   -1.271)
    1.610   -2.515

21   0.046   21   0.110   25   0.064   25   0.075
(0.005,   0.041)   (0.017,   0.093)   (0.036,   0.028)   (-0.200,   0.275)

0.253   -1.802   0.222   0.251   -6.043   0.264
(-0.166,   -1.635)   (-3.418,   -2.626)
    -3.678   -9.210

26   0.062   26   0.049
(0.029,   0.033)   (-0.330,   0.379)

0.540   -9.672   0.173
(-4.496,   -5.176)
c)   -11.432

Fig. 18. (Conclusion). The EDM localized to resemble most closely the populational MEC (external, also shown for comparison) : EDM (→ MEC). The modes are divided into subsets associated with constituent atoms in the $[V_2O_5]$ -cluster (reactant A, **Part a**), ring carbons (**Part b**), and hydrogen atoms (**Part c**) of toluene (reactant B), in the model toluene—$[V_2O_5]$ system. The isolated reactant charges (MNDO-toluene, scaled-INDO-cluster) have been used to generate the hardness tensor. The primary displaced atoms are identified by the largest open circle in the specified group of atoms, or it is identified by an asterisk (the $H_{(o)}$ case). In each two-diagram area, the first diagram represents the EDM localized on the MEC shown in the second diagram. The numerical data reported in the EDM plots are arranged in a way identical to that in Fig. 14. The numbers above each MEC diagram are $\phi_i \equiv (dN/dN_i)_{E=min} = \phi_i^A + \phi_i^B$, $\phi_i^A = (dN_A/dN_i)_{E=min}$, and $\phi_i^B = (dN_B/dN_i)_{E=min}$

Finally, let us examine the EDM ($\rightarrow$ IDM) and PNM resolutions of the AIM *in situ* FF indices (see Fig. 9A–c); they are reported graphically in Figs. 19 and 20, respectively. One concludes from these figures that the EDM ($\rightarrow$ IDM) set gives the most condensed collective-mode interpretation of the inter-reactant (*N*-constrained) CT, with only two unstable (negative mode hardness) modes, $\alpha = 1$ and 3, participating strongly in the toluene $\rightarrow$ cluster electron transfer; the two modes, $\beta = 2$ and 4, provide only a slight correction, to reproduce the positive feature of $f_{C(H_3)}^{CT}$, seen in Fig. 9A–c. The PNM set gives a resolution of comparable compactness, with modes $\gamma = 1$ and 3, also unstable, generating the gross features of $\mathbf{f}^{CT}$, and the modes $\gamma = 5, 8, 9,$ and 20 generating a slight correction, mainly producing negative values at the methyl hydrogens. It follows from a comparison of Figs. 16, 17 and 19, 20 that the most compact collective-mode resolution of $\mathbf{f}$, representing the open M as a whole, is obtained in the

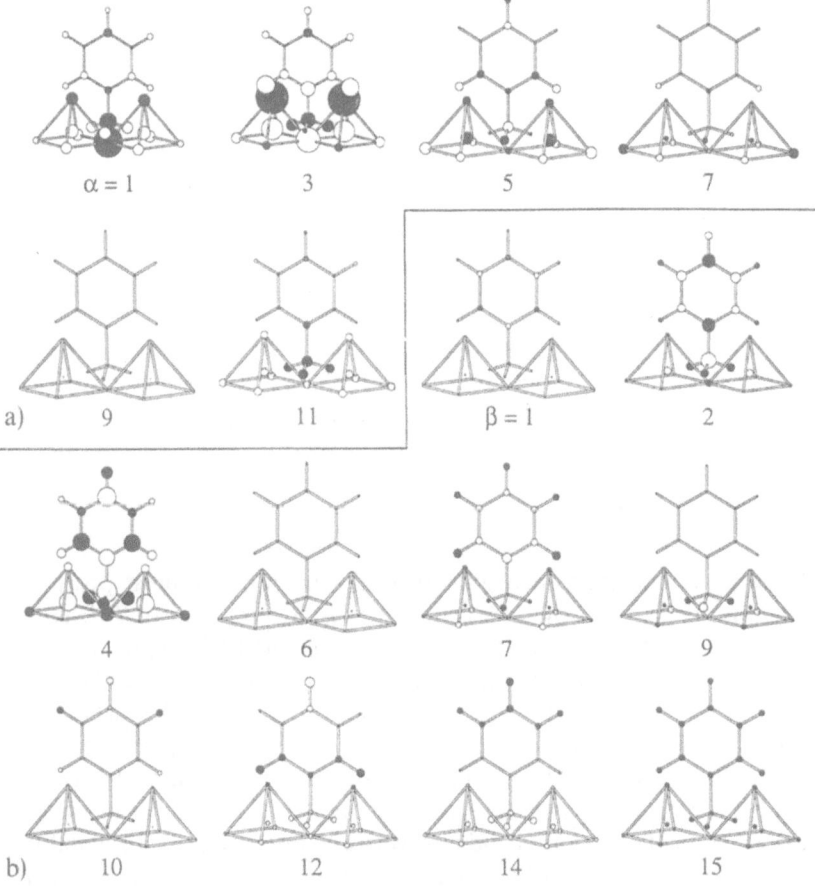

Fig. 19. The EDM ($\rightarrow$ IDM) resolution of the $\mathbf{f}^{CT}$ (see Figs. 9 A–c and 10). The diagrams represent the $\omega_{i,\alpha(\beta)}^{CT(EDM)}$ quantities, i = 1, ..., m, and $\alpha \in A, \beta \in B$; only the $F_{\alpha(\beta)} \neq 0$ modes are shown

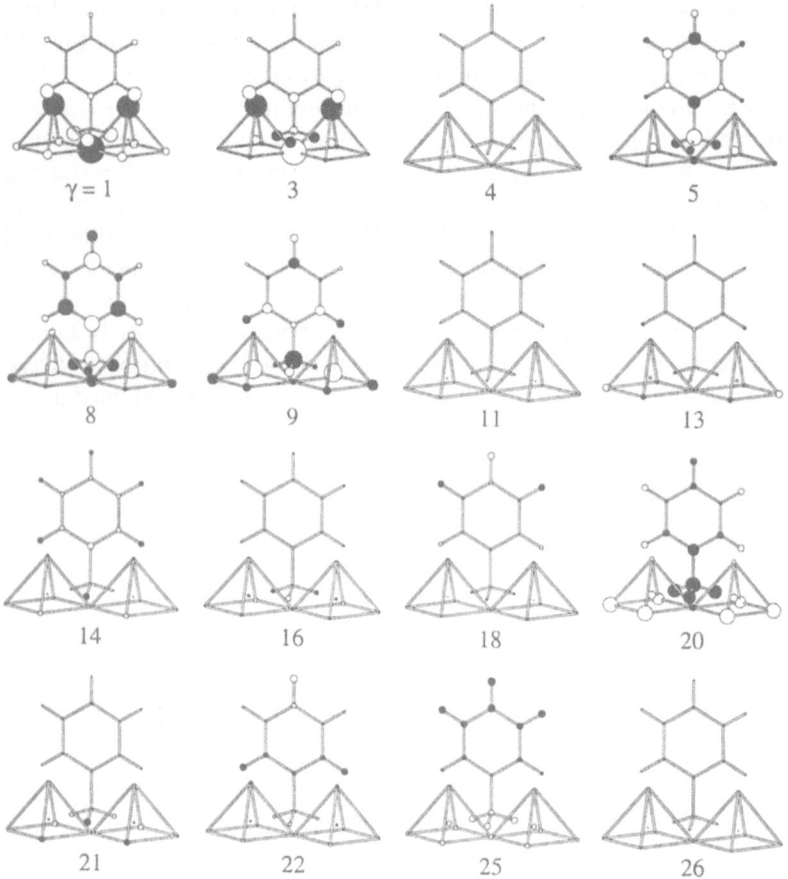

Fig. 20. The PNM resolution of f^{CT} (see Figs. 9 A–c, 12, and caption to Fig. 16)

PNM representation (total decoupling), while the N-constrained process of electron transfer between the interacting species, a resolution of f^{CT}, is best described in the EDM ($\rightarrow$ IDM) framework. Thus, the PNM represent "natural" collective displacements for an external CT involving hypothetical electron reservoir, while EDM ($\rightarrow$ IDM) are the "natural" reaction channels for the B $\rightarrow$ A CT, preserving a memory of the actual charge couplings between reactants. Hence, both sets are valuable interpretative and diagnostic tools, designed to provide the most condensed description of the charge rearrangement accompanying an external (between M and reservoir) and internal (between A and B in M) test CT. As such they may be called *"natural populational coordinates" in the theory of chemical reactivity.*

5.3 Inter-Reactant Coupling Modes

The IDM of Sect. 5.1 are defined as the eigenvectors of the block-diagonal hardness matrices:

$$\eta_i \equiv \begin{pmatrix} \eta^{A,A} & 0^{A,B} \\ 0^{B,A} & \eta^{B,B} \end{pmatrix} \quad \text{and} \quad \eta_i^{rel} \equiv \begin{pmatrix} \eta_{A,A}^M & 0^{A,B} \\ 0^{B,A} & \eta_{B,B}^M \end{pmatrix} \tag{217}$$

which neglect the respective inter-reactant charge coupling blocks, $\eta^{A,B}$ and $\eta_{A,B}^M$. The complementary hardness matrices, $\eta_e = \eta - \eta_i$ and $\eta_e^{rel} \equiv \eta^{rel} - \eta_i^{rel}$ are:

$$\eta_e \equiv \begin{pmatrix} 0^{A,A} & \eta^{A,B} \\ \eta^{B,A} & 0^{B,B} \end{pmatrix} \quad \text{and} \quad \eta_e^{rel} \equiv \begin{pmatrix} 0^{A,A} & \eta_{A,B}^M \\ \eta_{B,A}^M & 0^{B,B} \end{pmatrix} \tag{218}$$

Their eigenvectors, which we call the *inter-reactant modes* (IRM), provide yet another collective charge displacement reference frame, which can be used to describe the CT processes (internal or external) in a general reactive system $M = A - - - B$. For example, the IRM corresponding to the rigid matrix η_e, $\mathcal{U}^e = (\mathcal{U}_1 | \mathcal{U}_2 | \cdots | \mathcal{U}_m)$, are defined by the equation

$$\mathcal{U}_e^\dagger \eta_e \mathcal{U}_e = \hbar \quad \text{(diagonal)}, \quad \mathcal{U}_e \text{ (unitary)} \tag{219}$$

Let us examine the above eigenvalue problem in more detail. We explicitly separate the reactant components $\mathcal{U}_X (X = A, B)$ in the representative eigenvector $(\mathcal{U}_y)^\dagger \equiv \mathcal{U}^\dagger = (\mathcal{U}_A^\dagger, \mathcal{U}_B^\dagger)$; also, to simplify notation, we denote $\hbar \equiv \hbar_{y,y}$. Rewriting Eq. (219) gives:

$$\eta^{A,B} \mathcal{U}_B = \hbar \mathcal{U}_A \quad \text{and} \quad \eta^{B,A} \mathcal{U}_A = \hbar \mathcal{U}_B \tag{220}$$

Multiplying the first of Eqs. (220) from the left by $\eta^{B,A}$ and inverting the transformed equation yields the expression for $\mathcal{U}_B$ in terms of $\mathcal{U}_A$:

$$\mathcal{U}_B = \hbar G_B^{-1} \eta^{B,A} \mathcal{U}_A, \quad G_B = \eta^{B,A} \eta^{A,B} \tag{221}$$

which can be used to obtain from the second Eq. (220) the following eigenvalue equation for the $\mathcal{U}_A$ component:

$$A \mathcal{U}_A = \hbar^2 \mathcal{U}_A, \quad A = (\eta^{A,B} G_B^{-1} \eta^{B,A})^{-1} G_A \tag{222}$$

where $G_A = \eta^{A,B} \eta^{B,A}$ (see Eq. (221)). A similar elimination process for the $\mathcal{U}_B$ component gives the corresponding eigenvalue problem:

$$B \mathcal{U}_B = \hbar^2 \mathcal{U}_B, \quad B \equiv (\eta^{B,A} G_A^{-1} \eta^{A,B})^{-1} G_B \tag{223}$$

with the same eigenvalue as in Eq. (222).

Therefore, the $\mathcal{U}_A$ and $\mathcal{U}_B$ components of $\mathcal{U}$ are the eigenvectors of the $(n \times n)$–A and $[(m - n) \times (m - n)]$–B matrices, respectively, which exhibit the same eigenvalue, $\hbar^2$. Let us assume, e.g., that $m > 2n$. Hence, the n eigenvectors $\mathcal{U}_A$ of A will combine with their n conjugates $\mathcal{U}_B$ among the eigenvectors of B, which exhibit the same (positive) eigenvalue, whereas the remaining $m - 2n$ eigenvectors of $\mathbf{B}$, $\mathcal{U}'_B$, will combine with the zero component on A, giving rise to the *B-localized* (ℓ) IRM:

$$(\mathcal{U}'_\gamma)^\dagger = [\mathbf{0}^\dagger_A, (\mathcal{U}'_B)^\dagger], \quad \gamma = 2n + 1, \ldots, m \tag{224}$$

It can be quickly verified, using Eq. (219), that the eigenvalues of such localized modes vanish identically, $\hbar^\ell = 0$. Thus, the B-localized modes can only polarize B and they play no part in the CT between reactants. However, they can lead to a net change in N_B, thus exhibiting a nonvanishing external CT component of reactant B, i.e., generally nonvanishing w_γ and w_γ^B values, and $w_\gamma^A \equiv 0$. When the CT between reactants is considered these modes will only participate in polarizing B, thus generally exhibiting nonvanishing F_γ^{CT}, ω_γ^{CT}, and $\Omega_{B,\gamma}^{CT}$, and identically vanishing $\Omega_{A,\gamma}^{CT}$. The remaining $2n$ eigenvectors will be delocalized throughout M. They can be grouped in pairs of IRM exhibiting opposite eigenvalues, of the same magnitude, $\hbar = |\hbar|$ and $\hbar = -|\hbar|$:

$$\mathcal{U}(\pm|\hbar|) = [\mathcal{U}_A(\hbar^2) \pm \mathcal{U}_B(\hbar^2)]2^{-1/2}. \tag{225}$$

Again, one can verify the eigenvalues using Eq. (219).

There is a close analogy between this IRM representation and the localized molecular orbitals of the SCF MO theory. Namely, the qualitative localized MO's, bonding and anitbonding, can also be constructed by combining a pair of directed hybrid orbitals of the bonding atoms with the non-bonding hybrids describing lone electron pairs on the constituent atoms. Following this analogy, one could consider the above inter-reactant decoupling scheme as constructing the $\mathcal{U}_A$ and $\mathcal{U}_B$ charge-displacement "hybrids" of the reactants, that are designed to reflect the actual reactant interaction in M. As such they also belong to the category of *in situ* reactivity concepts.

In Fig. 21 we present the illustrative IRM diagrams for the toluene–$[V_2O_5]$ system, together with the relevant quantities: their eigenvalues, hardness parameters, and those quantities that reflect their participation in the external and internal (toluene $\rightarrow [V_2O_5]$) CT. The corresponding $\{w_{i,\gamma}\}$ and $\{\omega_{i,\gamma}^{CT}\}$ diagrams, providing the IRM resolution of the f and f^{CT} vectors, respectively, are shown in Figs. 22 and 23.

It follows from Fig. 22 that the external FF indices, f are determined in effect by the two complementary IRM, $\gamma = (6, 21)$, which originate from the same $\mathcal{U}_A$ and $\mathcal{U}_B$ components. Reference to Fig. 9A–e shows that their combined effect indeed correctly reproduces the cluster part of f, with only minor charge shifts on the constituent atoms of toluene. The toluene part of the FF diagram is seen to be mainly due to the background, nonselective mode, $\gamma = 26$. The four toluene-localized IRM, $\gamma = 12$–15 are seen to be only slightly involved in the external CT.

Let us now examine the corresponding IRM resolution of f^{CT} (Fig. 9A–c). In this case, the toluene-localized mode $\gamma = 12$ is used to diminish the amount of external CT going to the ortho-carbon. The negative f_i^{CT} feature of the terminal

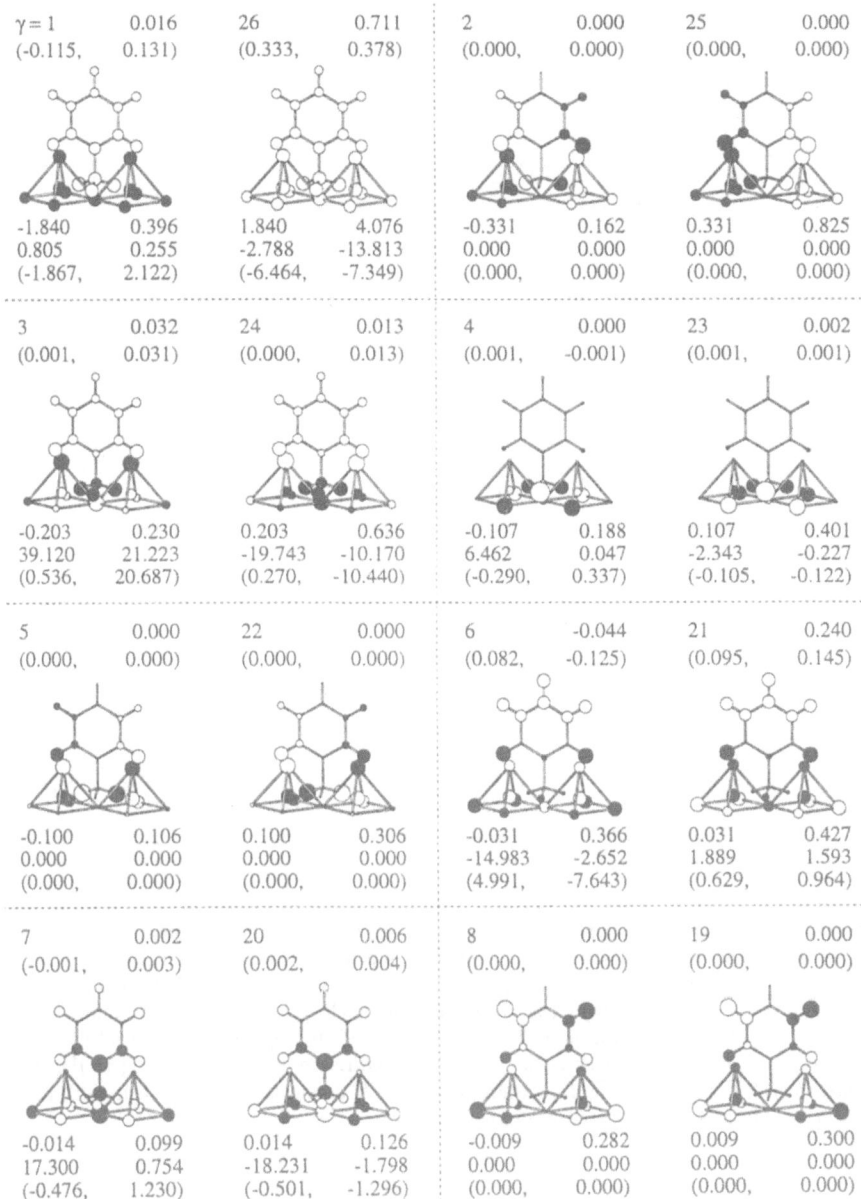

$\gamma = 1$	0.016	26	0.711	2	0.000	25	0.000
(-0.115,	0.131)	(0.333,	0.378)	(0.000,	0.000)	(0.000,	0.000)
-1.840	0.396	1.840	4.076	-0.331	0.162	0.331	0.825
0.805	0.255	-2.788	-13.813	0.000	0.000	0.000	0.000
(-1.867,	2.122)	(-6.464,	-7.349)	(0.000,	0.000)	(0.000,	0.000)
3	0.032	24	0.013	4	0.000	23	0.002
(0.001,	0.031)	(0.000,	0.013)	(0.001,	-0.001)	(0.001,	0.001)
-0.203	0.230	0.203	0.636	-0.107	0.188	0.107	0.401
39.120	21.223	-19.743	-10.170	6.462	0.047	-2.343	-0.227
(0.536,	20.687)	(0.270,	-10.440)	(-0.290,	0.337)	(-0.105,	-0.122)
5	0.000	22	0.000	6	-0.044	21	0.240
(0.000,	0.000)	(0.000,	0.000)	(0.082,	-0.125)	(0.095,	0.145)
-0.100	0.106	0.100	0.306	-0.031	0.366	0.031	0.427
0.000	0.000	0.000	0.000	-14.983	-2.652	1.889	1.593
(0.000,	0.000)	(0.000,	0.000)	(4.991,	-7.643)	(0.629,	0.964)
7	0.002	20	0.006	8	0.000	19	0.000
(-0.001,	0.003)	(0.002,	0.004)	(0.000,	0.000)	(0.000,	0.000)
-0.014	0.099	0.014	0.126	-0.009	0.282	0.009	0.300
17.300	0.754	-18.231	-1.798	0.000	0.000	0.000	0.000
(-0.476,	1.230)	(-0.501,	-1.296)	(0.000,	0.000)	(0.000,	0.000)

Fig. 21. (Begining).

vanadyl oxygens, $O_{(1)}$, is seen to be reproduced by the common $\mathcal{U}_A$ component of the complementary modes, $\gamma = 3$ and 24, while that of the bridging oxygen, $O_{(2)}$, is due to the two cluster components of modes $\gamma = (7, 20)$ and $\gamma = (9, 18)$.

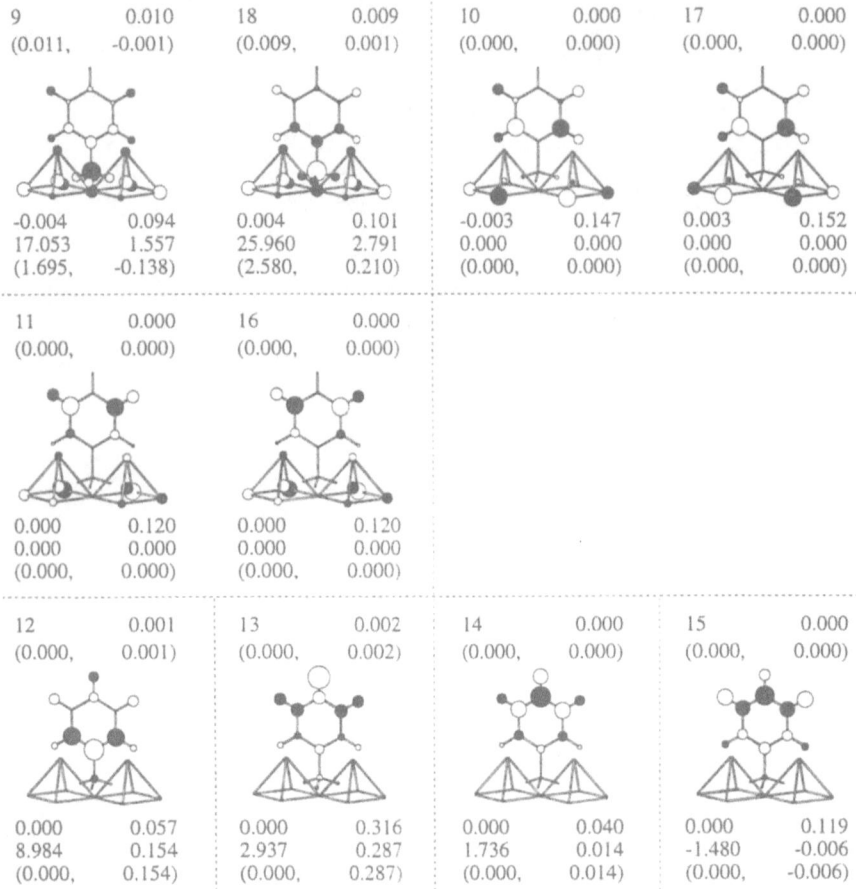

Fig. 21. (Continued). The IRM for the toluene–[V_2O_5-cluster] system; isolated reactant charges have been used. The delocalized modes are grouped in pairs of the (γ, m − γ + 1) IRM corresponding to the same magnitude of the eigenvalue, and the B-localized modes, $\gamma = 12$–15 are shown in the last row of the figure. The numerical data above the mode diagram include: the mode number (ordered according to increasing $\hbar_{\gamma,\gamma}$), w_γ (first row), and its reactant resolution into (w_γ^A, w_γ^B) (second row). Below each diagram the following quantities are listed: $\hbar_{\gamma,\gamma}$, $\tilde{\eta}_{\gamma,\gamma} \equiv [\mathscr{U}^e]^\dagger \eta \mathscr{U}^e]_{\gamma,\gamma}$ (first row), F_γ^{CT}, ω_γ^{CT}, and ($\Omega_{A,\gamma}^{CT}$, $\Omega_{B,\gamma}^{CT}$) (third row). The CT parameters have been calculated for the assumed toluene → [V_2O_5] electron transfer

We therefore conclude that the IRM, solely determined by reactant interaction, are well suited to describe CT processes. They provide the most compact description yet of such charge reorganizations in molecular systems. Their attractiveness is increased by the fact, that the IRM decoupling process identifies well defined collective reactive and non-reactive components of reactants. This has far reaching implications for model catalytic systems involving large surface clusters, much exceeding the adsorbate in size. In such structures the localized

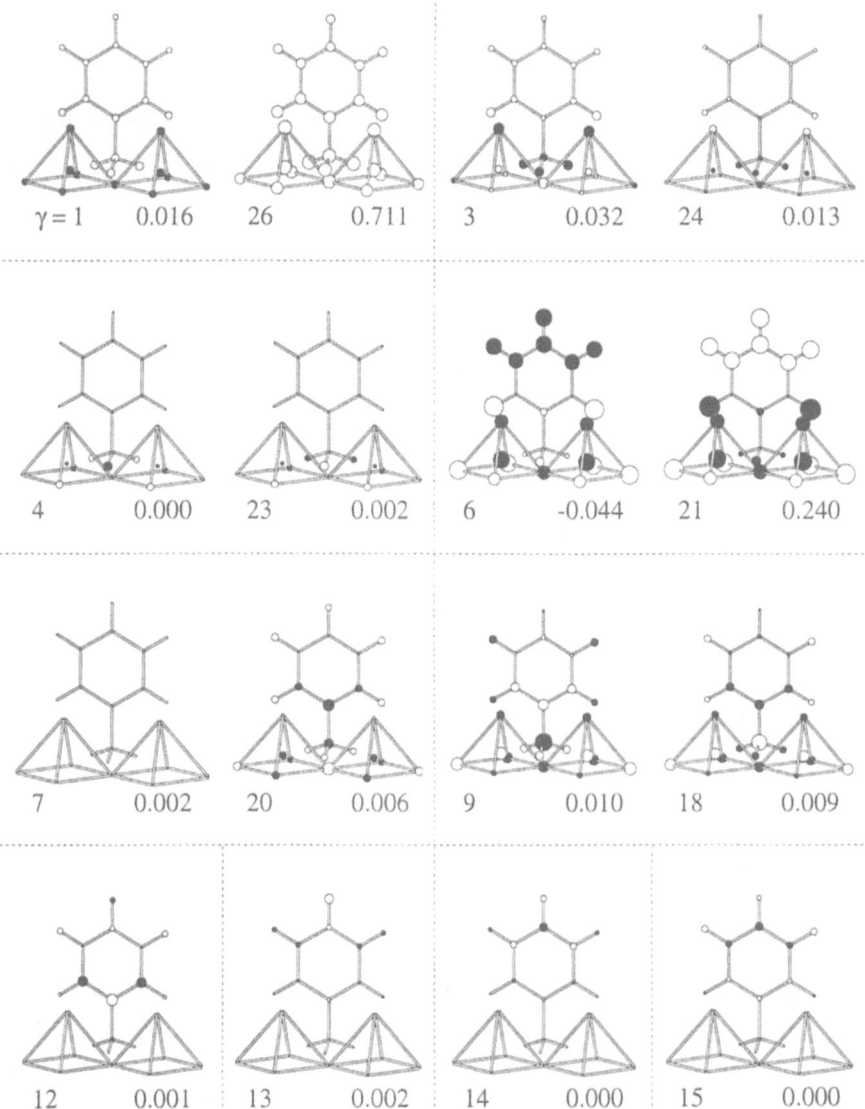

Fig. 22. The IRM resolution, $\{w_{i,\gamma}^{(IRM)}\}$, of the FF diagram of Fig. 9 A–e (see captions to Figs. 10, 16, and 17)

(environmental) modes will solely involve the substrate, whereas the delocalized (reactive) modes will involve both the cluster and the substrate. The substrate-localized modes, which substantially contribute to f or f^{CT} could then be used to diagnose the influence of the cluster's environment on the active site of the surface or on the adsorbate.

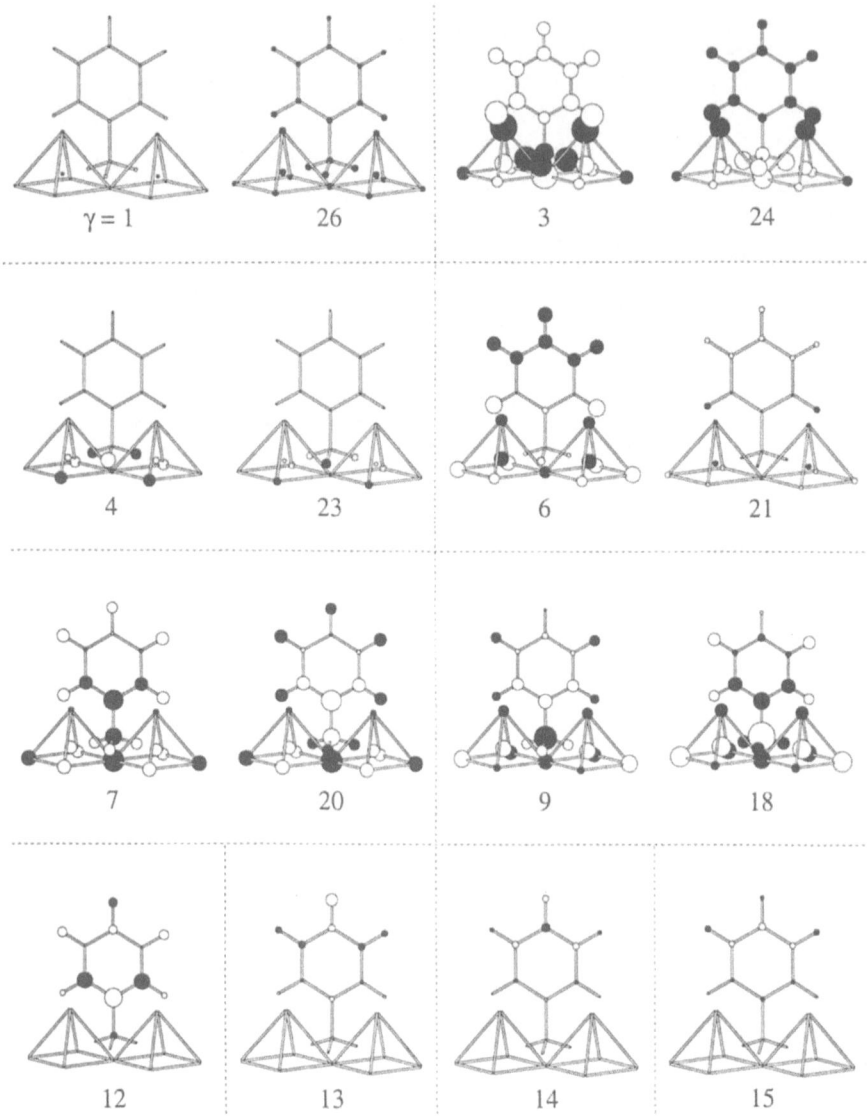

Fig. 23. The IRM resolution, $\{\omega_{i,\gamma}^{CT(IRM)}\}$, of the f^{CT} diagram of Fig. 9 A–c (see caption to Fig. 10)

5.4 Stability Considerations

In this section we summarize charge stability criteria and their physical implications [8, 25, 47]. For reasons of clarity we consider a general partitioning of the system under consideration, $\mathscr{S}$, into two complementary subsystems, $\mathscr{S}_1$ and

$\mathscr{S}_2$. We shall examine the external and internal cases separately. The former refers to the combined system, consisting of M and a hypothetical electron reservoir, (M!r), while the latter corresponds to the closed M, $M = (\mathscr{S}_1!\mathscr{S}_2)$. In the N-constrained case the complementary subsystems denote, e.g., the A and B reactants in a general acceptor-donor reactive system, the adsorbate and substrate in heterogenous catalysis, etc. Let $N_1^{\mathscr{S}}$ and $N_2^{\mathscr{S}}$ denote the respective numbers of electrons, grouped in the vector $\mathbf{N}_{\mathscr{S}} = (N_1^{\mathscr{S}}, N_2^{\mathscr{S}})$, with $N = N_1^{\mathscr{S}} + N_2^{\mathscr{S}}$ denoting the global number of electrons in $\mathscr{S}$. The relevant condensed hardness tensor in the subsystem resolution, $\mathbf{\eta}^{\mathscr{S}}(1|2)$, is defined by Eqs. (35) and (44) in the relaxed and rigid approximations, respectively. Its elements define the susbsystem's effective hardnesses, $\eta_1^{\mathscr{S}} = \eta_{1,1}^{\mathscr{S}} - \eta_{1,2}^{\mathscr{S}}$, etc., $\eta_{CT}^{\mathscr{S}} = \eta_1^{\mathscr{S}} + \eta_2^{\mathscr{S}}$, and the related response properties, e.g., the FF indices, $f_1^{\mathscr{S}} = \eta_2^{\mathscr{S}}/\eta_{CT}^{\mathscr{S}}$ and $f_2^{\mathscr{S}} = \eta_1^{\mathscr{S}}/\eta_{CT}^{\mathscr{S}}$ of the subsystems.

Let us assume internal equilibrium in $\mathscr{S}$, which corresponds to the mutually open subsystems, $\mathscr{S}^\circ = (\mathscr{S}_1!\mathscr{S}_2)$, with equalized chemical potentials, $\mu_1^{\mathscr{S}} = \mu_2^{\mathscr{S}} = \mu_{\mathscr{S}} \equiv \partial E/\partial N$, at the global chemical potential level. The internal stability refers to intra-$\mathscr{S}$ (hypothetical) charge displacements, $d\mathbf{N}_{\mathscr{S}}(\Delta) = (\Delta, -\Delta)$, that preserve N. The corresponding quadratic energy change due to this polarizational displacement from the initial internal equilibrium state:

$$dE(\Delta) = (\mu_1^{\mathscr{S}} - \mu_2^{\mathscr{S}})\Delta + \frac{1}{2}\eta_{CT}^{\mathscr{S}}\Delta^2 = \frac{1}{2}\eta_{CT}^{\mathscr{S}}\Delta^2 \equiv d^2 E(\Delta) \tag{226}$$

implies the following internal stability criterion, $d^2E(\Delta) > 0$:

$$\eta_{CT}^{\mathscr{S}} > 0 \quad \text{or} \quad a \equiv (\eta_{1,1}^{\mathscr{S}} + \eta_{2,2}^{\mathscr{S}})/2 > \eta_{1,2}^{\mathscr{S}} \tag{227}$$

Let us similarly consider $\mathscr{S}$ to be in equilibrium with an external electron reservoir, characterized by its chemical potential μ_r, $(\mathscr{S}_1!\mathscr{S}_2!r)$, which implies the chemical potential equalization; $\mu_1^{\mathscr{S}} = \mu_2^{\mathscr{S}} = \mu_{\mathscr{S}} = \mu_r$. The virtual flow of electrons between $\mathscr{S}$ and r, $dN = -dN_r$, gives rise to the associated quadratic energy change:

$$dE(dN) = (\mu_{\mathscr{S}} - \mu_r)\,dN + \frac{1}{2}\eta_{\mathscr{S}}(dN)^2 = \frac{1}{2}\eta_{\mathscr{S}}(dN)^2 \equiv d^2 E(dN) \tag{228}$$

where the second-order energy change due to the infinitely soft (macroscopic) reservoir is neglected, and the global hardness $\eta_{\mathscr{S}}$ is:

$$\eta_{\mathscr{S}} = \partial\mu/\partial N = [\eta_{1,1}^{\mathscr{S}}\eta_{2,2}^{\mathscr{S}} - (\eta_{1,2}^{\mathscr{S}})^2]/\eta_{CT}^{\mathscr{S}} \tag{229}$$

Thus, the external stability criterion $dE(dN) > 0$ implies for the internally stable charge distribution (Eq. 227):

$$g^2 \equiv \eta_{1,1}^{\mathscr{S}}\eta_{2,2}^{\mathscr{S}} > (\eta_{1,2}^{\mathscr{S}})^2 \tag{230}$$

and this inequality is reversed for the internally unstable ($\eta_{CT}^{\mathscr{S}} < 0$) charge distribution in $\mathscr{S}$. Moreover, since

$$|g| = [4a^2 - (\eta_{2,2}^{\mathscr{S}} - \eta_{1,1}^{\mathscr{S}})^2]^{1/2}/2 \leq |a| \tag{231}$$

one obtains the following stability regimes for the typical region of coupling hardness between subsystems, $0 < \eta_{1,2}^{\mathscr{S}} < \eta_{1,1}^{\mathscr{S}}$ (we denote $\mathscr{S}_1$ as the harder subsystem, $\eta_{1,1}^{\mathscr{S}} \geq \eta_{2,2}^{\mathscr{S}}$):

$$0 \leq \eta_{1,2}^{\mathscr{S}} < g, \quad \text{internally (I) and externally (E) stable } \mathscr{S} : (+, +);$$

$$g \leq \eta_{1,2}^{\mathscr{S}} < a, \quad \text{I-stable and E unstable } \mathscr{S} : (+, -); \tag{232}$$

$$a \leq \eta_{1,2}^{\mathscr{S}} < \eta_{1,1}^{\mathscr{S}}, \quad \text{I-unstable and E-stable } \mathscr{S} : (-, +).$$

These are shown schematically in Fig. 24a.

Clearly, since the FF indices of the two subsystems are closely related to the structure of the condensed hardness tensor $\boldsymbol{\eta}^{\mathscr{S}}(1|2)$, these stability/instability regions can also be identified in terms of the $f_1^{\mathscr{S}}$ and $f_2^{\mathscr{S}}$ subsystem FF indices defining the orientation of the associated FF vector in the populational space of the two subsystems [25].

These qualitative conclusions have important consequences for catalytic systems [47]. Let us assume, for example, that $\mathscr{S} = $ (adsorbate|substrate), with

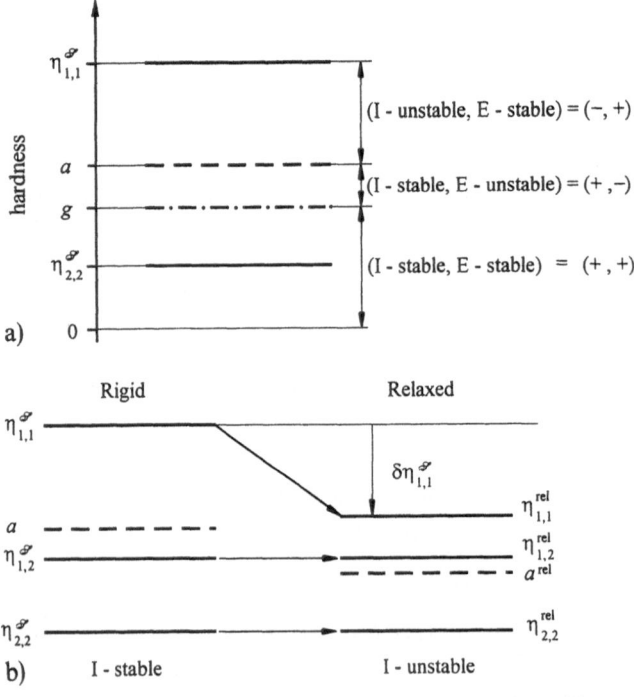

Fig. 24. (a) Internal (I) and external (E) stability ($+$) and instability ($-$) regions in $\mathscr{S} = (\mathscr{S}_1|\mathscr{S}_2)$ in terms of the coupling hardness $\eta_{1,2}^{\mathscr{S}}$ of $\mathscr{S} = (\mathscr{S}_1|\mathscr{S}_2)$; (b) the softening effect of the substrate ($\mathscr{S}_2$) upon the adsorbate ($\mathscr{S}_1$) in catalytic system $\mathscr{S}$, generating the internal charge instability

a large substrate representing the soft reactant ($\mathscr{S}_2$) and a small adsorbate corresponding to the hard reactant ($\mathscr{S}_1$). Since the substrate exerts a strong softening influence on the adsorbate (large, negative relaxational correction to $\eta^{\mathscr{S}}_{1,1}$ due to $\mathscr{S}_2$), $\delta\eta^{\mathscr{S}}_{1,1}(\mathscr{S})$ (see Fig. 24b), while the opposite effect of the adsorbate on $\eta^{\mathscr{S}}_{2,2}$ and on the coupling hardness $\eta_{1,2}$ are negligible and exactly vanishing, respectively, this implies a lowering of both the a and g levels relative to the fixed coupling hardness, eventually generating instabilities in the system. This mechanism is illustrated in Fig. 24b. Thus, in this CSA, phenomenological perspective, the catalyst acts as a generator of the I- and/or E-instabilities through its softening influence upon the adsorbate. The instability conditions correspond to reactive charge distributions, which spontaneously readjust

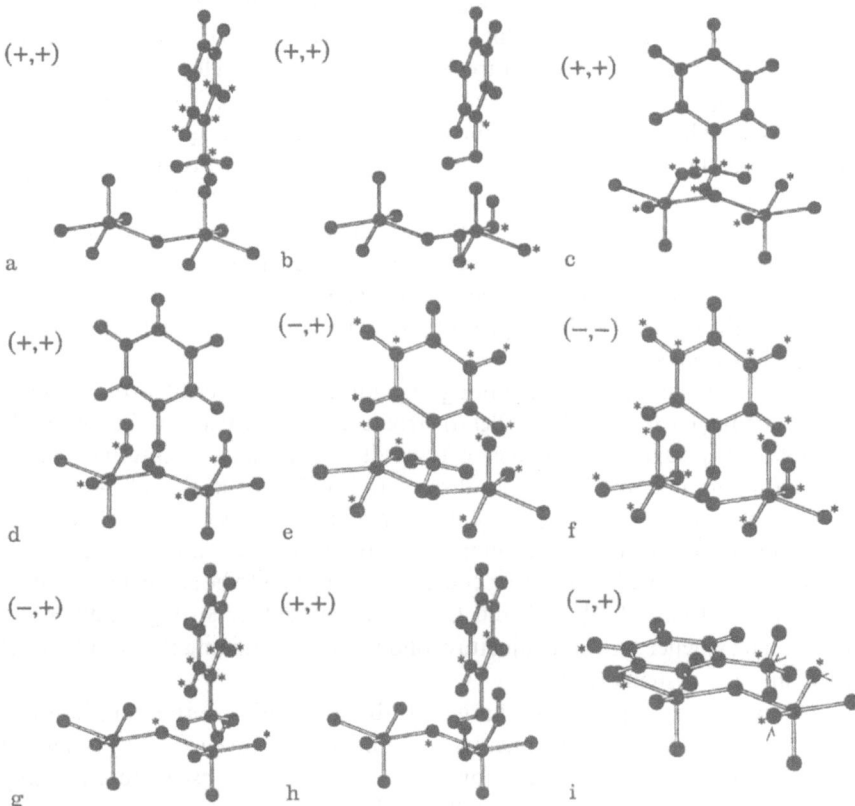

Fig. 25. Alternative toluene chemisorption systems involving the bipyramidal surface cluster of vanadium pentoxide. The structures (**a, c, e, g, i**) represent molecularly adsorbed toluene, while the remaining systems (**b, d, f, h**) model the dissociative adsorption, with two methyl hydrogens chemically bonded to the surface oxygens at the pyramid bases. Asterisks indicate positions which give rise to unstable MEC. The three atoms of the parallel complex (**i**), marked with an arrow, denote the extra instabilities appearing when the closed-system constraint ($dN = 0$) is imposed. The results are taken from Ref. 8. At each diagram the (I, E) stability diagnosis is also indicated (see Fig. 24)

themselves through both the intra-adsorbate and the adsorbate-substrate charge flows, eventually leading to adsorbate reconstruction and/or desorption (dissociative or molecular).

Consider now the illustrative example of alternative toluene-chemisorption systems shown in Fig. 25 [8], involving a bipyramidal cluster modelling the V_2O_5 "active site" of the (010)-surface shown in Fig. 26a. The molecularly adsorbed structures (Panels a, c, e, g, i) correspond to the scaled INDO-optimized distances between the "frozen" geometries of the adsorbate and the active site [44]. They include the perpendicular molecular adsorptions of toluene, through its methyl group, to the vanadyl (terminal, singly coordinated to vanadium atom) oxygen $O_{(1)}$ (Panel a), to the bridging (doubly coordinated to the vanadium atoms) oxygen $O_{(2)}$ (Panels c and e), and to the vanadium atom (Panel g), as well as the parallel adsorption of the benzene ring to the pyramid bases (Panel i). The remaining structures represent hypothetical dissociated forms (Panels b, d, f, and h), in which the two hydrogens of the methyl group are already bonded to the oxygens at the base of the pyramid. The rigid structure of the remaining toluene fragment has been assumed [44]. We have used the scaled INDO charges to generate the condensed hardnesses $\eta^{\mathscr{S}}$ (toluene|cluster) to diagnose the (I, E)-stabilities of these model chemisorption systems. The predicted diagnoses are reported in Fig. 25. We would like to stress, that in this model of the "active site" the size of the V_2O_5 cluster and toluene are comparable, so that the automatic classification: the soft (substrate) and hard (toluene) reactants, no longer applies. When diagnosing the dissociatively adsorbed structures, the dissociated hydrogens of the "activated" structures (b, d, f, h) have been included in toluene.

It follows from Fig. 25 that, among the molecular adsorption arrangements, the perpendicular (e, g) and parallel (i) structures are predicted to be internally unstable. This suggests a tendency towards a spontaneous internal charge reconstruction, possibly leading to dissociation of bonds in the adsorbate. On the basis of SCF MO studies [44] of these systems, one expects the CH bonds of the methyl groups to be affected most in the perpendicular cases; in structure e, the $H_{(o)}$ atoms of the benzene ring can also be expected to be strongly activated. In the parallel case, both CH and CC bonds should be strongly influenced. Additional evidence for this conjecture follows from examining the populational MEC in these systems [8, 27].

The only case of predicted charge instabilities, both external and internal, among the dissociated perpendicular structures is observed for the system of Panel f; this may suggest a need for a further charge reconstruction, possibly involving the cluster's environment. The MEC arguments [8, 27] also indicate that the CH ring bonds in the ortho-position relative to the methyl group are strongly affected by the vanadyl oxygens, besides the methyl bonds already partly dissociated in Panel f.

In Table 1 we report the condensed hardness data for complementary subsystems in a much larger cluster of the V_2O_5 crystal consisting of 126 atoms, shown in Fig. 26b. They illustrate a softening influence on the diagonal hardness

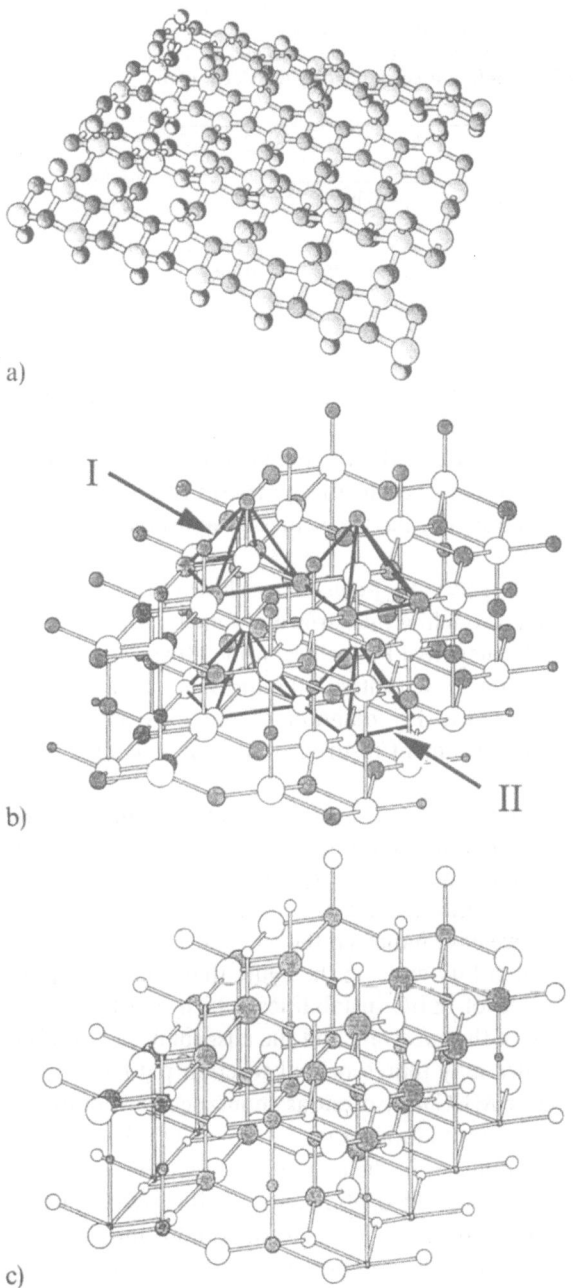

Fig. 26. Perspective view of the (010)-surface of the vanadium pentoxide with different mutual arrangements of neighboring pyramidal [VO_5] units (**Panel a**). **Panels b** and **c** correspond to neutral, stoichiometric surface cluster including two layers of the (010) surface pyramids (126 atoms); in **Panel b**, which illustrates the SINDO AIM net-charge distribution, the bipyramidal subsystems I and II are shown (see Table 1); **Panel c** represents the AIM FF distribution diagram

Table 1. Rigid and relaxed condensed hardnesses (a.u.) of the complementary subsystems X and Y of the stoichiometric, two-(010)-layer cluster of vanadium pentoxide including 126 atoms (Fig. 26); here X denotes the bipyramidal model of the surface active site (I or II) and Y stands for the corresponding remainder of the cluster: $[V_2O_5] = (X|Y)$.

X	$\eta_{X,X}$		$\eta_{Y,Y}$		$\eta_{X,Y}$
Fig. 26b	rigid	relaxed	rigid	relaxed	
I	0.1206	0.0698	0.0681	0.0680	0.0661
II	0.1246	0.0753	0.0679	0.0675	0.0664

of a small subsystem X (bipyramidal units shown in Fig. 26b), due to its interaction with its large complementary subsystem Y. Notice that the relaxational softening changes the structure of the rigid condensed hardness matrix $\boldsymbol{\eta}^{\mathscr{S}}(X|Y)$, producing a rather soft (close to instability) charge distribution, with the off-diagonal, coupling hardness $\eta_{X,Y}^{\mathscr{S}}$ having values comparable to the two relaxed diagonal hardnesses.

The AIM FF indices of the large $[V_2O_5]$-cluster are shown in Fig. 26c. It is seen that in the first (0 1 0)-layer, modeling the surface of Panel a, the positively charged vanadium atoms exhibit negative FF values, while all surface oxygens (negatively charged) are characterized by positive FF indices, marking relatively soft electron distributions. Thus, the FF diagram of the bipyramidal unit changes when the effect of its environment in the crystal is included (compare Figs. 8b and 26c).

The external MEC can be used to test the system's minimum energy responses to a localized, test reduction/oxidation. Both the sign of the coordinate's hardness parameter Γ_k and the mode topology can be of interest in diagnosing the system's reactivity patterns and its sensitivity to environmental changes [7, 8, 27, 28]. In Fig. 25 we have indicated the AIM positions, k, defining unstable external MEC ($\Gamma_k < 0$). It follows from the figure that each adsorption pattern has its own distinct area of MEC instabilities, which can be used to diagnose the mutual activation of the substrate and the adsorbate as a result of adsorption. Such unstable modes may represent real reaction pathways, since displacements along the unstable MEC may occur as spontaneous responses to fluctuations in the AIM electron populations. One also observes that such an "activation" region generally diminishes in the assumed dissociative structures, relative to the corresponding molecular adsorption cases. This is particularly so in pairs of related panels: (c, d) and (g, h). Such an overall diminishing trend of the MEC instability regions provides *a posteriori* validation of the assumed dissociation pathway of the chemisorption system. An exception to this rule is seen in Panels (e, f) of the figure; this may suggest that for this adsorption arrangement the postulated dissociation pattern has missed other, energetically more favourable, pathways, e.g., those involving the toluene *ortho*-hydrogens, or that there is still a strong trend towards geometry relaxation in the assumed activated structure. Reference to Panels a and b shows that, after dissociating two methyl hydrogens, the area of MEC instability inside the toluene fragment is limited to a single ring carbon (bonded to the CH fragment of the methyl group). One also observes that the three pyramid-base oxygens have been activated as a result of toluene dissociation; this indicates a possibility of subsequent dissociations of the lattice hydroxyls or oxygens. It follows from the figure that perpendicular adsorption on the bridging oxygen selectively "activates" only the methyl group: in all remaining cases different positions in the benzene ring are also predicted to give rise to unstable MEC.

In all perpendicular arrangements, the instability areas from the external and internal MEC have been found to be identical. Therefore, they are dominated by internal polarization displacements, with little effects of an external CT on the predicted instability (activation) areas. In the parallel arrangement i, however, the internal MEC give rise to a wider instability region, which includes the methyl carbon and the neighboring lattice oxygens, in addition to the *para*-hydrogen and its near-neighbour lattice oxygen; the latter is predicted to give rise to unstable external MEC. Thus, the external CT may moderate the internal, purely polarizational MEC instabilities by an intervention of the compensating charge flow involving the environment of the system.

6 Illustrative Application to Large Model Catalytic Systems

6.1 Water-Rutile System

This section is devoted to the CSA study of the molecular/dissociative adsorption of water on the (1 1 0)-surface of rutile [7, 48], in which we employ some of the concepts of the preceding sections. The mechanism for these processes has recently been formulated by Kurtz et al. [49], on the basis of synchrotron radiation resonant photoemission data and complementary spectroscopic information. It has been established [50] that, at room temperature, H_2O is adsorbed molecularly on the (1 0 0)-surface, and dissociatively on the (1 1 0)-surface (see Fig. 27). On the nearly-perfect surface, water adsorbs on a five-fold O-coordinated Ti site, and interacts with a bridging O ligand; a subsequent bending of this singly coordinated water towards the bridging oxygen represents the reaction coordinate for the dissociation of water [48] (see Fig. 28). On the defective surface, the preferred adsorption site is the bridging-O vacancy, which allows double coordination of water (see Fig. 28); the water dissociation reaction coordinate would require a rotation of the chemisorbed molecule towards the neighboring bridging oxygens, which needs a much larger activation energy than the bending of the singly coordinated water [48]. In both cases two OH species are created as a result of the water dissociation. The previous study of the H_2O - - - (1 1 0)-rutile system [48] has involved the non-stoichiometric, 50-atom,

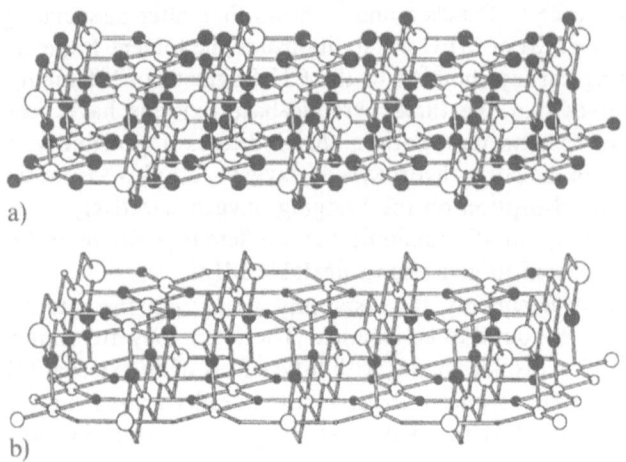

a)

b)

Fig. 27. (Begining).

rutile cluster. Here we have extended the cluster representation of the (1 1 0)-sur-face to a full 2-layer stoichiometric model, consisting of 147 atoms, as shown in Fig. 27. The input AIM charges (Panel a) have been determined from semiem-pirical SINDO1 SCF MO calculations [51]. They generate the AIM FF indices f, shown in Panel b of the figure, a resolution of which into the main CT-active PNM of Panel c is displayed in Panel d of the figure. The seven modes included carry about 96% of the external CT.

It follows from Fig. 27 that it is the softest PNM, $\gamma = 1$, which exhibits the strongest influence on the AIM charges, although most of the external CT is due to the non-selective hardest mode $\gamma = 147$. As would be expected, the atoms at the cluster boundary are predicted to be the softest (Panel b), due to the unsaturated valences in this region. When one adopts the cluster approximation one is automatically faced with the question of its adequacy. Clearly, such a treatment is justified, when the perturbation due to adsorption (in the active site region) remains strongly localized in the region of the cluster center and rapidly decays with increasing distance from the surface bond. Also, in an adequate cluster representation, hypothetical charge displacements at the boundary of the cluster should have negligible influence upon the chemisorption region. Both these aspects can be quickly tested using the MEC [8, 28]. This is explicitly demonstrated in Fig. 29, where diagrams of selected MEC for the rutile cluster of Fig. 27 are reported. The first diagram, with the primary displacement localized on the titanium surface site, shows that the test reduction on this atom creates AIM shifts, the range of which does not extend appreciably beyond the second neighbors of the primarily displaced atom. This *a posteriori* validates the small cluster size used in the previous study. It also shows an effective screening of this primary perturbation through a polarization of the environment nearest the displaced site. The second diagram indicates, that the

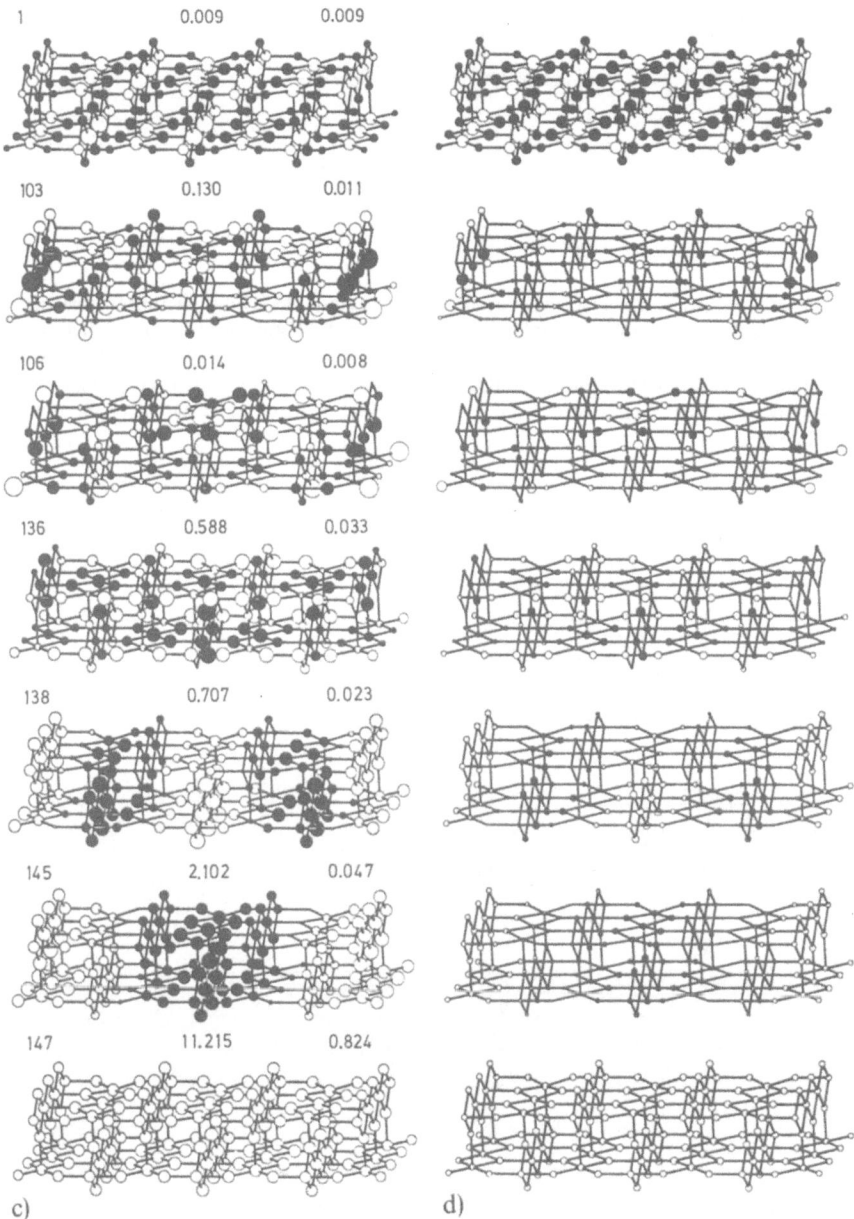

Fig. 27. (Continued). The two (110)-layer, 147 atom rutile cluster charges (**Panel a**), AIM FF indices *f* (**Panel b**), the most important external CT-active PNM (**Panel c**), and their contributions to *f*, $\{w_{i,\gamma}\}$ (**Panel d**). The numbers above each diagram in **Panel c** are: the mode number, principal hardness (ordering criterion), and w_γ

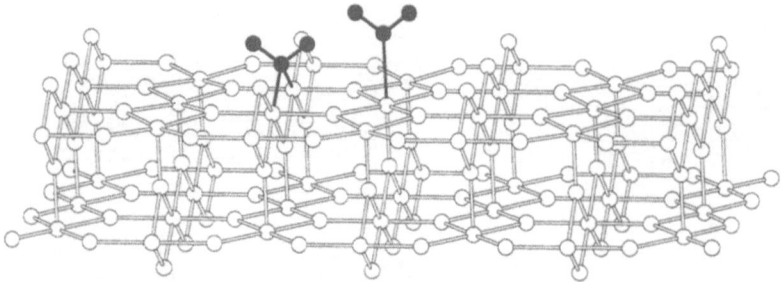

Fig. 28. Alternatives sites of a molecular adsorption of water on the (110)-rutile surface cluster (146 atoms) on the central, five fold O-coordinated Ti site (the singly coordinated water molecule) and on the bridging oxygen vacancy (the doubly coordinated water molecule). The corresponding reaction coordinates of the dissociative adsorption that produce two OH species involve the bending of water towards the bridging oxygen of the singly coordinated adsorbate and rotation of the doubly coordinated water

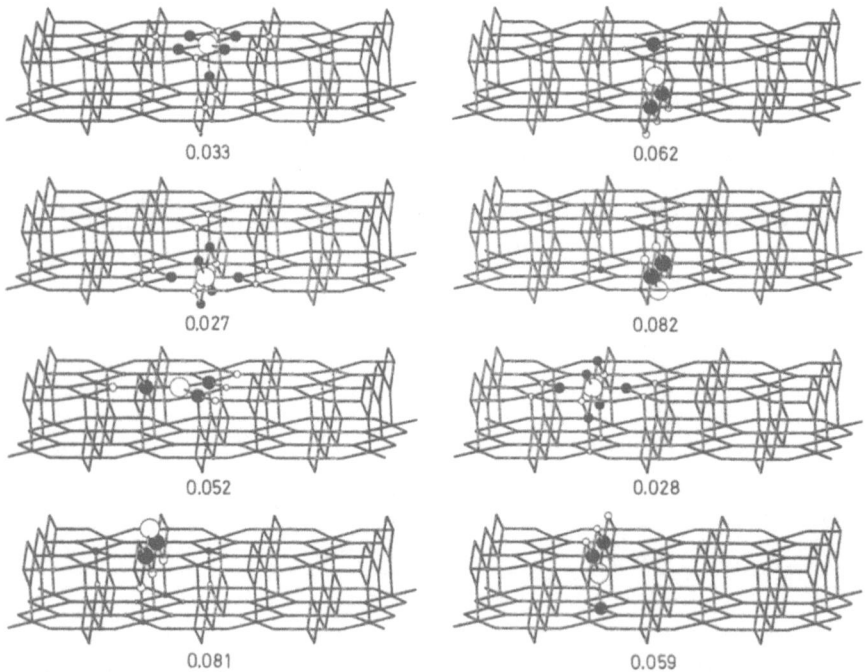

Fig. 29. Representative external MEC diagrams for a model (147 atom) cluster of the (110)-rutile surface. The numbers below the diagrams report Γ_k values (a.u.). The primarily displaced atom k is identified by the largest open circle. Only displacements around the central five-fold coordinated Ti site of the surface are included in the figure

valence state of the triply coordinated oxygen atoms that link the two layers exerts a strong influence on the five-fold O-coordinated titanium site, so it is vital to include both this atom and its nearest neighbors in the cluster that models this preferred active site. A similar conclusion follows from the last diagram of the figure, which reflects the influence of these layer-bridging $O_{(3)}$ atoms on both the surface bridging $O_{(2)}$ atoms and their neighbors, e.g., the six-fold O-coordinated Ti atoms. As seen in the second row diagrams, the test charge displacements on the second-layer atoms, in the vicinity of the central titanium site, have practically no influence upon this site. The third row MEC illustrate the range of the remaining perturbations, at locations close to the central active site. They show that the test reductions on the surface $O_{(3)}$ atoms tend to generate charge response on the near surface atoms and have only a minor effect upon the second-layer and the layer-bridging atoms. All these observations support the adequacy of the cluster of Fig. 27, and of the previous choice of its 50-atom subsystem [48] for modelling the active site of the nearly perfect (110)-surface.

Let us now examine the FF quantities for the molecular adsorption cluster (Fig. 30 A) and the transition-state (Fig. 30 B) along the water dissociation reaction coordinate. These chemisorption clusters exhibit the bridging-$O_{(2)}$ vacancy in the vicinity of the $Ti_{(5)}$ adsorption site.

A comparison of Figs. 27 b, 30 A-a, and 30 B-a shows that the AIM FF indices of the cluster change very little as a result of the water molecular adsorption, both in the equilibrium and bent structures, and of the removal of the missing oxygen. In all these diagrams the titanium atoms are seen to be the soft lattice units that accept most of the inflowing electrons from the reservoir and from the triply-coordinated lattice oxygens; the negative FF indices of the $O_{(3)}$ atoms indicate that they exhibit equilibrium charge displacements opposite to those of the system as a whole. The bridging $O_{(2)}$ sites are predicted to remain practically unaffected by the external CT. These predictions are the opposite of those resulting from the previous calculations on a small, non-stoichiometric rutile cluster [48].

The remaining panels of Fig. 30 b, c, and d, correspond to closed chemisorption systems, representing the diagonal, off-diagonal, and total CT FF indices, respectively, for water → rutile electron transfer.

Let us first examine the molecular adsorption of Part A. The resultant charge response pattern of Panel d is seen to result mainly from the diagonal part on water (Panel b) and the off-diagonal part on the cluster (Panel c). In contrast to the f distribution of Panel a, the f^{CT} quantities are strongly localized at the adsorption site and its close vicinity. The phases of the charge displacements in Panel d show that this internal CT is mainly from the water oxygen to the Ti site of the cluster, as expected during formation of the coordination bond; however, the observed accompanying charge adjustments on the water hydrogens that diminish bond polarity relative to that in the isolated water molecule, should lengthen the O-H bonds, in accordance with the mapping relations of Sect. 2.3. Thus, one again observes in Panels b and c, that the adsorbate-

A   B

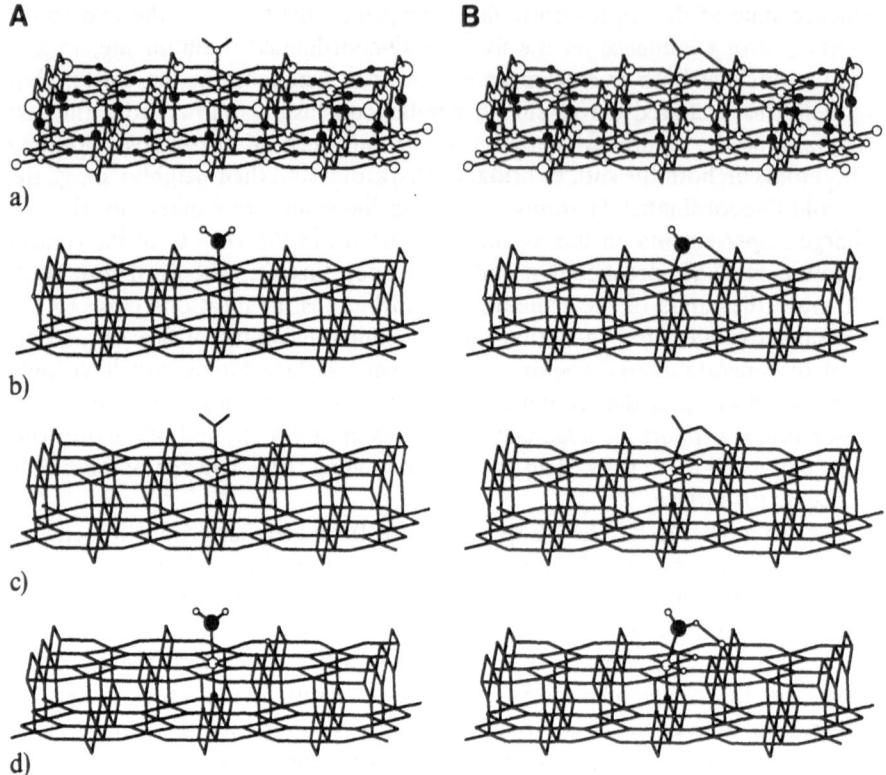

a)

b)

c)

d)

Fig. 30. The AIM FF indices for the singly coordinated water chemisorption complex of Fig. 28 (**Part A**) and the corresponding transition-state (**Part B**) toward the water dissociation (with a tilt angle of $21°$ relative to the structure of Fig. 28). The **Panels a** and **d** correspond to f (open chemisorption cluster) and f^{CT} (closed chemisorption cluster), respectively, while **Panels b** and **c** represent the diagonal and off-diagonal AIM contributions to the resultant f^{CT} of **Panel d**. All CT quantities have been calculated for the assumed water → rutile electron transfer. See also the caption to Fig. 9

substrate interaction promotes both reactants, so that the negative feature of the donor atom of the basic reactant (water) faces the positive feature of the acceptor atom of the acidic reactant (rutile cluster), therefore generating a covalent (coordination) bond.

The f^{CT} quantities for the transition-state structure of Part B are seen to be very similar to those reported in Part A; however, the symmetry-breaking due to the adsorbate bending has created an increased off-diagonal response on the bridging oxygen, a clear sign of partial formation of the new OH bond.

The water-rutile cluster system provides a good example for testing the utility of the IRM representation of Section 5. 3, since only three pairs of reactive complementary modes involve water, the remaining 143 environmental modes represent only the internal charge redistribution in the substrate and thus take no part in CT between the reactants. The reactive IRM for the molecular and transition-state adsorption cases are displayed in Parts A and B of Fig. 31, respectively, together with their external and internal CT and hardness characteristics. Their participation in f (Panels a) and f^{CT} (Panels b) is illustrated by the diagrams of Parts A and B of Fig. 32.

Consider first the molecular adsorption structure. In this case, the essentially antisymmetric modes $\gamma = (2, 148)$ (notice the symmetry breaking due to the oxygen vacancy on the surface) do not participate in charge displacement between water and rutile (see Figs. 31 A and 32 A), so that the CT-reactivity information is basically limited to the four reactive IRM which originate from the two components on each reactant. However, in the transition-state struc-

A

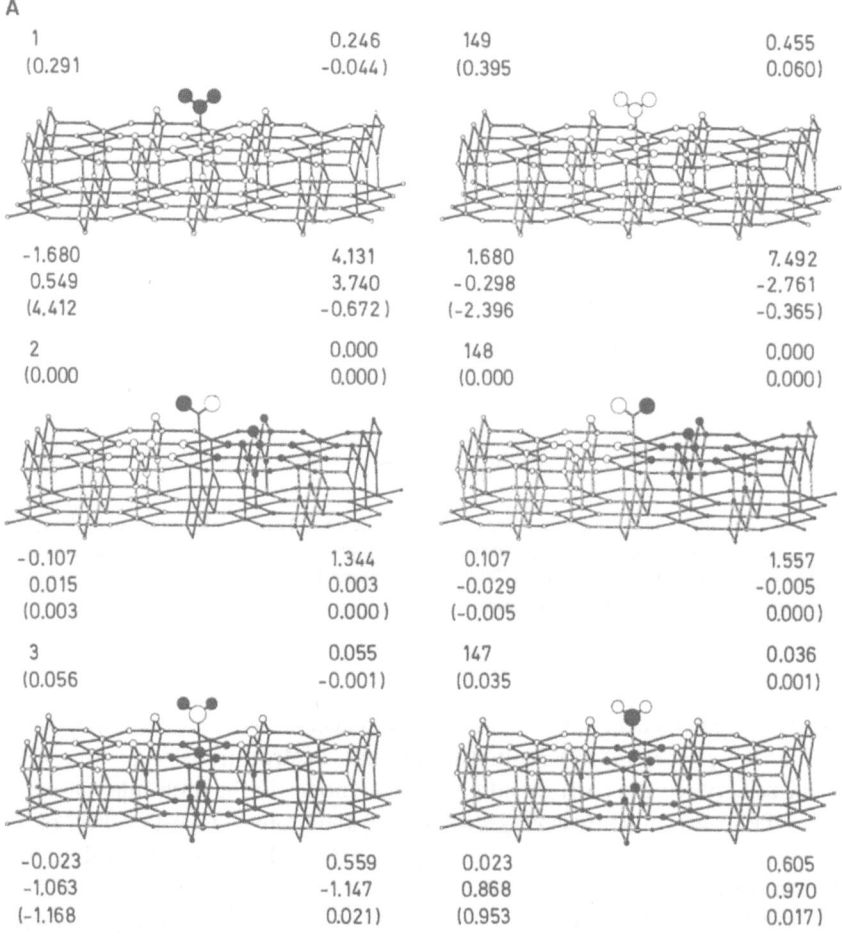

| 1 | 0.246 | 149 | 0.455 |
| (0.291 | -0.044) | (0.395 | 0.060) |

-1.680	4.131	1.680	7.492
0.549	3.740	-0.298	-2.761
(4.412	-0.672)	(-2.396	-0.365)

| 2 | 0.000 | 148 | 0.000 |
| (0.000 | 0.000) | (0.000 | 0.000) |

-0.107	1.344	0.107	1.557
0.015	0.003	-0.029	-0.005
(0.003	0.000)	(-0.005	0.000)

| 3 | 0.055 | 147 | 0.036 |
| (0.056 | -0.001) | (0.035 | 0.001) |

-0.023	0.559	0.023	0.605
-1.063	-1.147	0.868	0.970
(-1.168	0.021)	(0.953	0.017)

Fig. 31. (Begining).

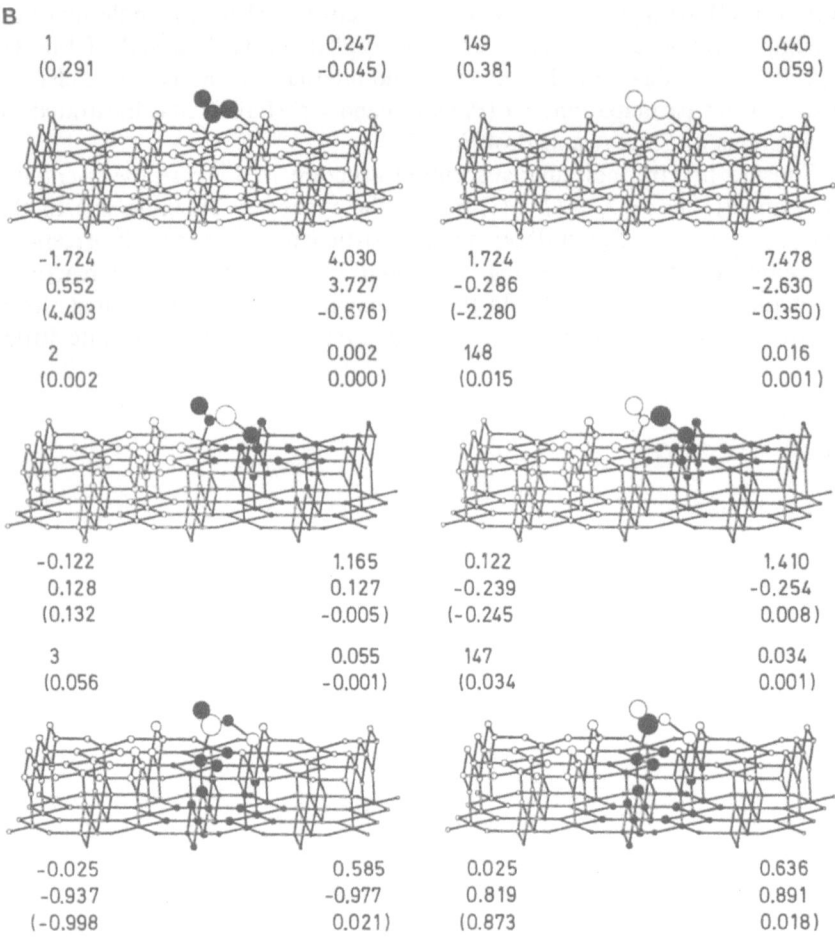

B

1	0.247	149	0.440
(0.291	-0.045)	(0.381	0.059)
-1.724	4.030	1.724	7.478
0.552	3.727	-0.286	-2.630
(4.403	-0.676)	(-2.280	-0.350)
2	0.002	148	0.016
(0.002	0.000)	(0.015	0.001)
-0.122	1.165	0.122	1.410
0.128	0.127	-0.239	-0.254
(0.132	-0.005)	(-0.245	0.008)
3	0.055	147	0.034
(0.056	-0.001)	(0.034	0.001)
-0.025	0.585	0.025	0.636
-0.937	-0.977	0.819	0.891
(-0.998	0.021)	(0.873	0.018)

Fig. 31. (Continued). The complementary reactive (delocalized) IRM for the molecular adsorption (**Part A**) and transition-state (**Part B**) chemisorption complexes of Fig. 30. The numerical data are reported in the same order as in Fig. 21. The CT parameters have been calculated for the assumed water (base, B) → rutile (acid, A) electron transfer

ture, with enhaced symmetry breaking, one detects relatively strong participation of the $\gamma = (2, 148)$ IRM in the internal electron-transfer process. The similarity between the corresponding reactive modes in the molecular case and that of the transition-state is very strong indeed, but one detects the effect of the new H–O bond being formed in Figs. 31 B and 32 B. The mode $\gamma = 1$ describes electron transfer from water to the cluster, with almost uniform distribution of the charge among the constituent atoms of each reactant. Its complementary mode, $\gamma = 149$, again with constant background displacements on each reactant, does not differentiate between local sites on both reactants either. The truly

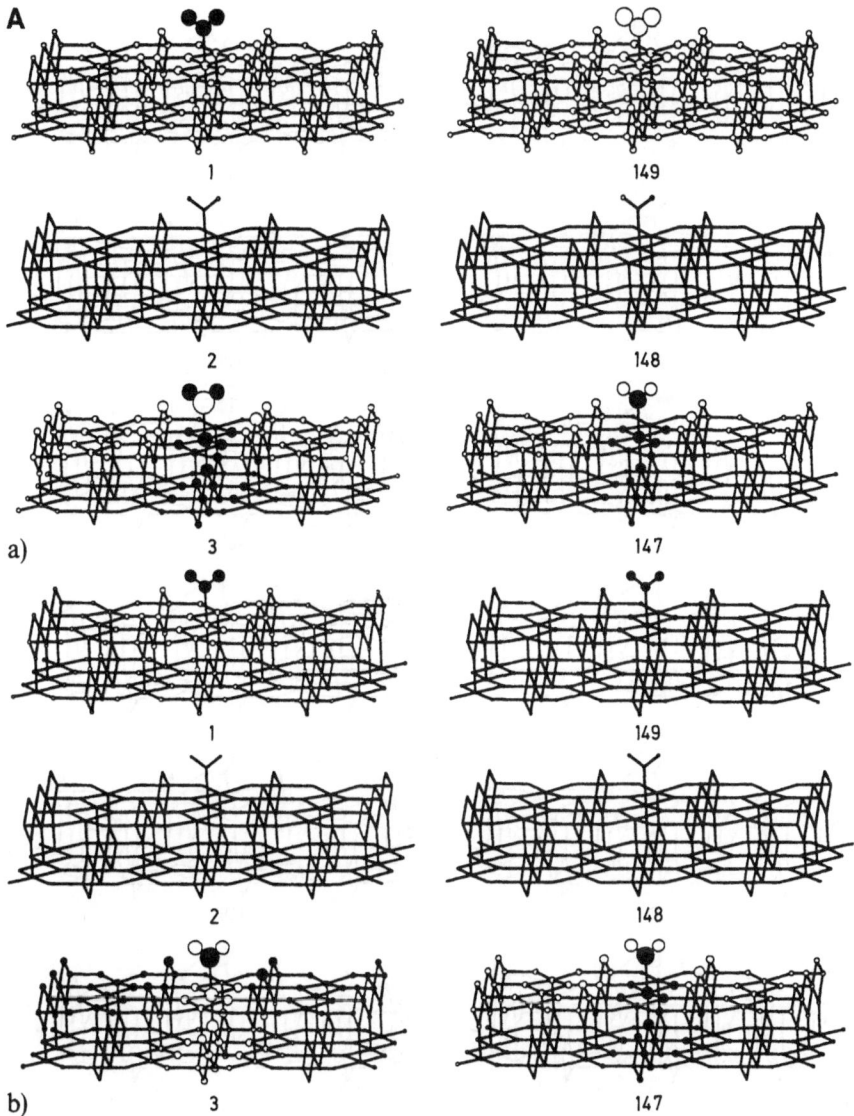

Fig. 32. (Begining).

reactive channels γ = (3, 147) generate the same reversal of the original bond polarity in the adsorbed water as that seen in the f^{CT} diagrams of Fig. 30 A. References to Fig. 32 A-b shows that the combined effect of these two modes is mainly limited to the water molecule, since the cluster components approximately cancel each other. The same cancellation is detected for their transition-state analogs in Fig. 32 B-b, where this activation of the water molecule is seen to be strengthened by modes γ = (2, 148). The external CT information on

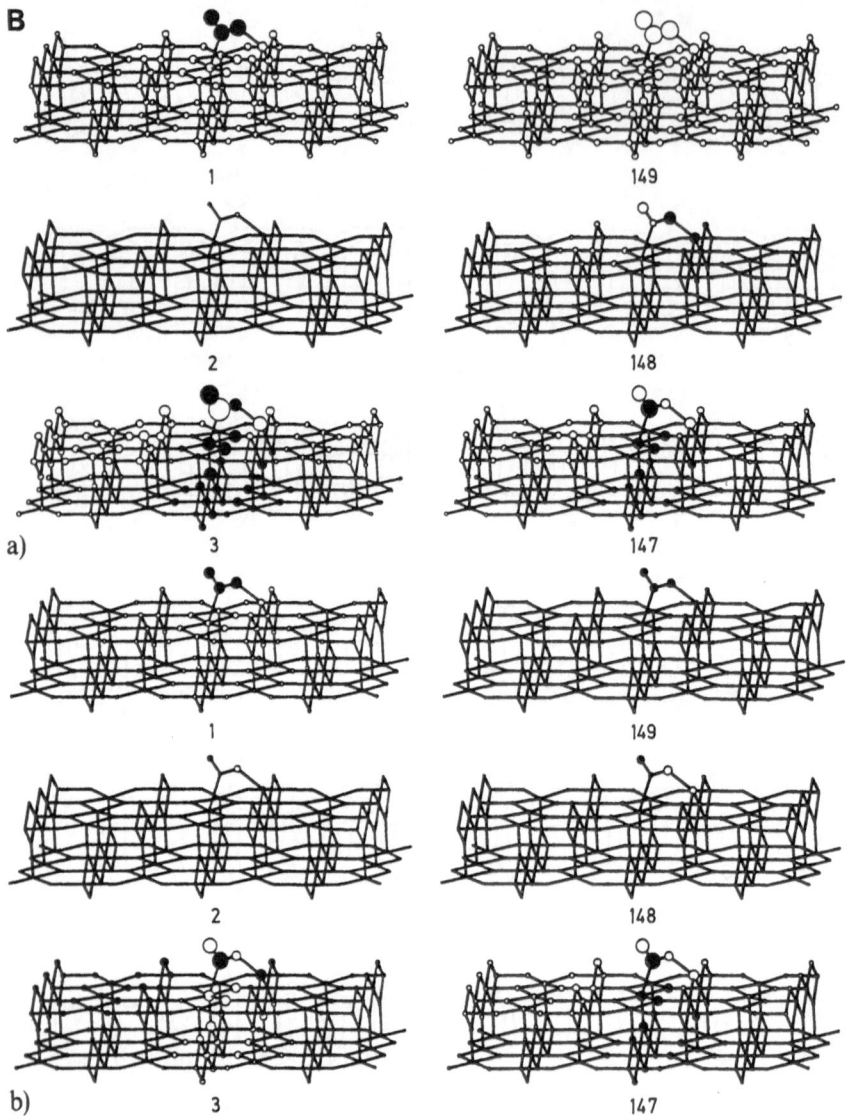

Fig. 32. (Continued). The reactive IRM (see Fig. 31) contributions to $f(\{w_{i,\gamma}^{(IRM)}\}$, **Panels a**) and $f^{CT}(\{\omega_{i,\gamma}^{(IRM)}\}$, **Panels b**) of the molecularly adsorbed (**Part A**) and the transition-state (**Part B**) chemisorption systems of Fig. 30

Panels a in Figs. 31 and 32 shows the moderating effect of the electron reservoir that is missing in the internal CT quantities. A comparison between Panels a and b of Fig. 32 indicates that the patterns of reactive IRM contributions to f and f^{CT} are very similar, except for the mode $\gamma = 148$ (Fig. 32 B), where the internal and external CT displacements exhibit opposite phases.

These illustrative results clearly demonstrate the great utility of the IRM reference frame in diagnosing charge transfer effects in chemisorption systems. They provide truly two-reactant reactive channels, the shapes of which resemble the interaction region of the corresponding FF indices. They have been found to be much more selective indicators than the corresponding external FF indices.

6.2 Toluene - [V_2O_5] System

The previous CSA, SCF MO and DFT studies [26–28, 44, 45, 76–78] of the vanadium oxide clusters and the chemisorption systems with toluene, have usually adopted a very small, strongly non-stoichiometric cluster representation of the surface active site, which included a very few idealized V_2O_5 pyramids of the surface layer. In a recent study [79] a relatively large, two-layer, nearly-stoichiometric cluster $V_{36}O_{98}$ shown in Fig. 33 has been used to examine the influence of both the surface size and of the supporting layer, in a truly two-reactant description of charge responses. The probing perpendicular and parallel structures of Fig. 33 generate the FF diagrams shown in Fig. 34, which compares the total *in situ* indices (Panels c), their diagonal (Panels a) and off-diagonal (Panels b) components, as well as the global FF diagrams characterizing the whole chemisorption system in contact with an external electron reservoir (Panels d).

It follows from Fig. 34 that the inclusion of the adsorbate-substrate charge couplings, which are responsible for the off-diagonal FF component and the relaxational changes in the diagonal FF component, is vital for adequate probing of chemical reactivity trends in large systems. The *in situ* FF indices are strongly localized in the chemisorption region, in contrast to the global FF diagrams; this localization validates the cluster approximation used in this extended analysis *a posteriori*. Namely, the range of the AIM populational displacements in the cluster, due to the inter-reactant CT, is seen to be limited to the close vicinity of the adsorption site. In all cases, the role of the supporting layer is rather small due to a relatively weak inter-layer charge coupling; in the parallel adsorption arrangements it is seen to be more pronounced.

In all cases the overall *in situ* AIM FF pattern is approximately the sum of the diagonal FF indices of toluene and the off-diagonal FF indices of the cluster. A comparison between Figs. 34 A, B and 9 A, B, respectively, shows that the *in situ* FF indices are very sensitive to the cluster representation and the adsorbate-substrate separation.

In Fig. 34 A-c, the methyl group hydrogens coordinate to both $O_{(1)}$ and $O_{(2)}$ surface oxygens. Formation of partial ($O_{(1)}$, $O_{(2)}$) ← $H_{(o)}$ bonds implies an effective removal of electrons from the $C_{(o)}$–$H_{(o)}$ bond region and from the methyl C–H bonds. Therefore, both the methyl and *ortho* ring carbons should be activated in this adsorption arrangement, thus becoming relatively susceptible to oxidation. The off-diagonal component of Fig. 34 A-b exhibits the dominating cluster charge relaxation trends. In the vicinity of the adsorption site, one

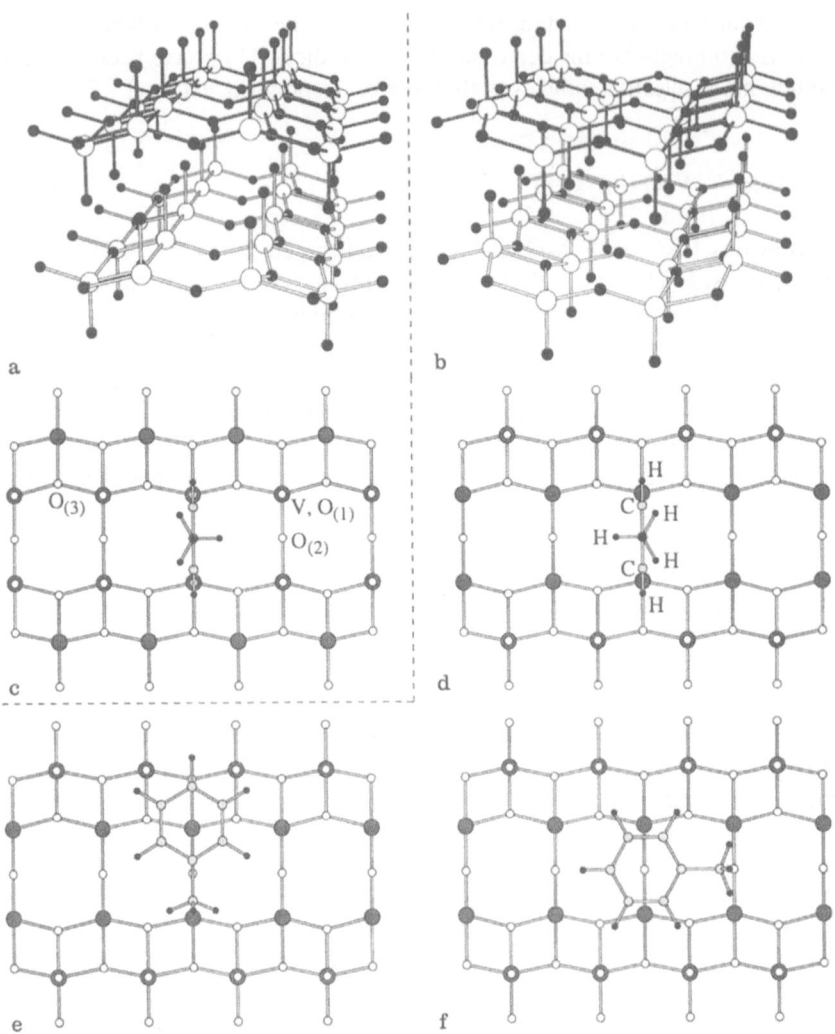

Fig. 33. Perspective views from the top-layer side (**Panel a**) and from the bottom-layer side (**Panel b**) of the representative, nearly-stoichiometric surface cluster $V_{36}O_{98}$. The geometric structure of the V_2O_5 (010)- surface cluster (see Fig. 26) involves the SINDO1 optimized intra-layer bond lengths [28] and the crystallographic values of the bond angles and the inter-layer V-O bond length. The surface layer views of alternative perpendicular adsorptions of toluene on the bridging oxygen $O_{(2)}$ from the top layer side (**c**) and from the bottom layer side (**d**), and two alternative parallel adsorptions (**Panels e, f**) on the bottom layer. In **Panel d** the same adsorbate-substrate separation as in Figs. 7a, 9 and 25e has been adopted, while in the remaining panels this separation has been increased by 1 Å relative to those shown in Fig. 25

A

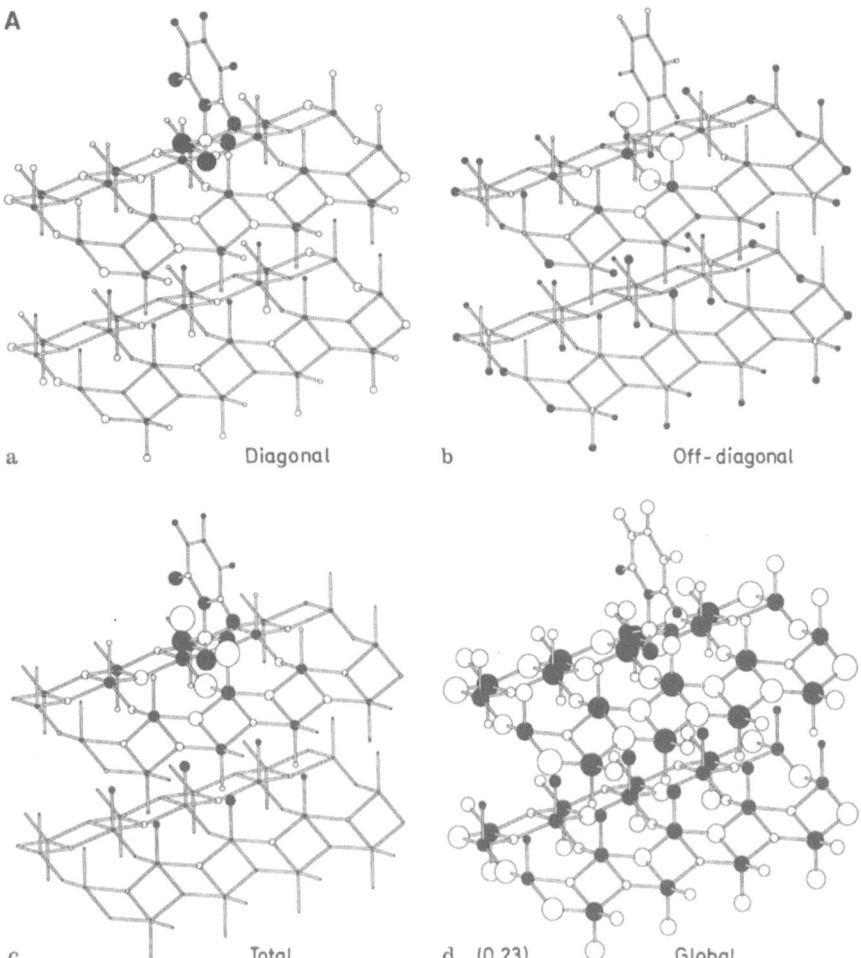

a   Diagonal   b   Off-diagonal

c   Total   d (0.23)   Global

Fig. 34. (Begining).

detects a slight reversal of the original O → V bond polarization, which weakens these cluster bonds; this may facilitate a subsequent participation of the active site oxygens in a selective oxidation of toluene. A comparison of Panel c and d in Fig. 34 A shows how relatively non-selective are the global FF reactivity indices, in comparison to their *in situ* analogs.

In the second perpendicular adsorption of Fig. 34 B only the methyl CH bonds of toluene and the bridging oxygen of the cluster are strongly activated, thus indeed indicating a preference towards a selective oxidation of the methyl group. The overall polarizational matching of the AIM charge shifts in the chemisorption region, with the donor atoms of the basic reactant (toluene)

B

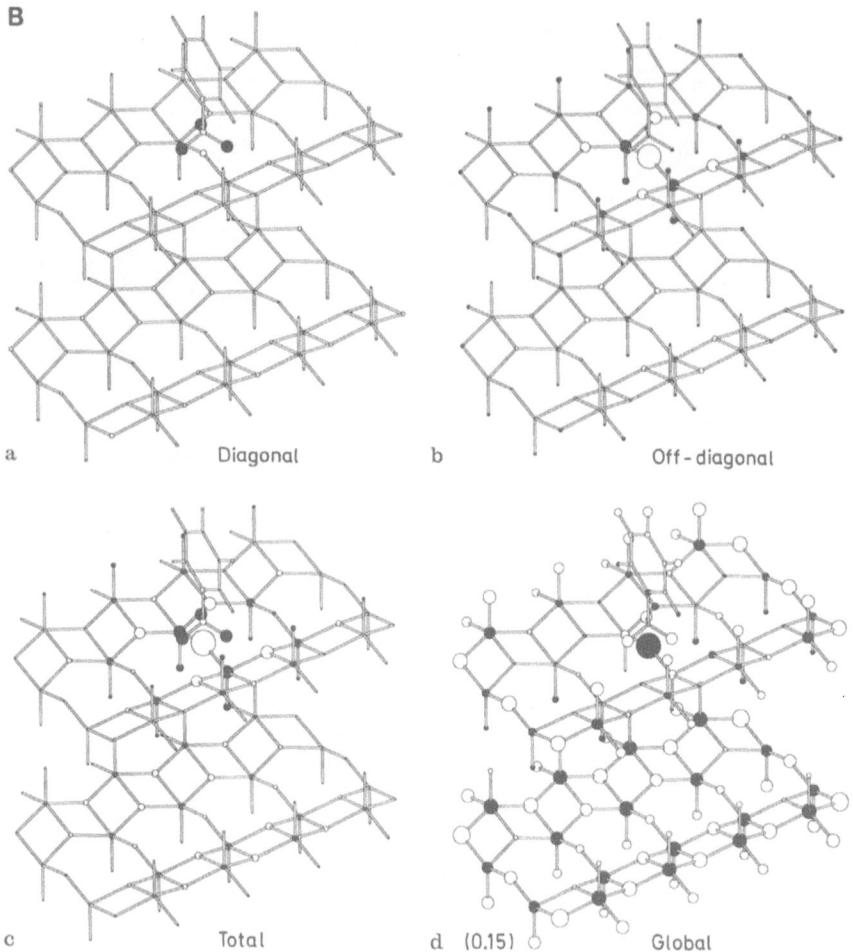

a   Diagonal   b   Off - diagonal

c   Total   d (0.15)   Global

Fig. 34. (Continued).

facing the acceptor atom of the acidic reactant (cluster), indicate a relatively soft charge rearrangement.

In both parallel arrangements (Figs. 34 C, D) the electrons are removed throughout the toluene; this indicates a tendency towards the adsorbate destructive oxidation. The lattice oxygens are strongly "released" (less bonded by vanadium atoms) due to the off-diagonal FF component; thus they may play a significant role in oxidizing the toluene fragments. This strong lattice reconstruction effect is also observed in the supporting layer, which modifies the primarily induced charge shifts of the surface layer.

For a more complete CSA study of the toluene-$[V_2O_5]$ system the reader is referred to the "Note added in proof" and ref. [79], while the allyl-$[MoO_3]$ system is examined in more detail in ref. [82].

C

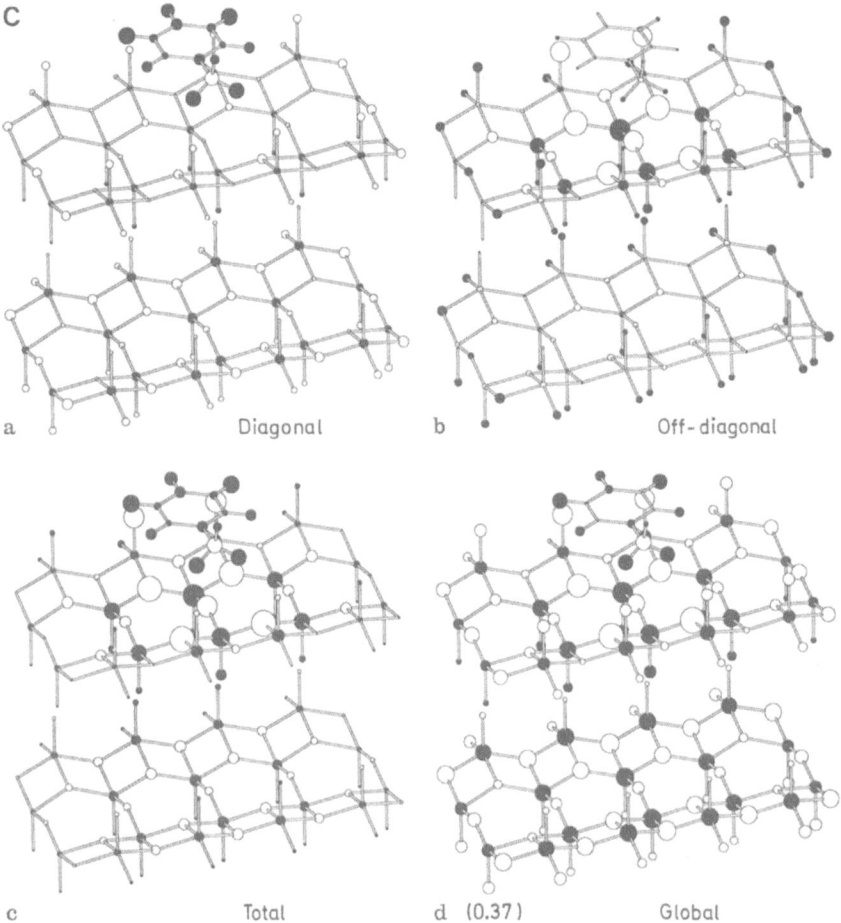

a   Diagonal   b   Off-diagonal

c   Total   d (0.37)   Global

Fig. 34. (Continued).

7 Conclusions and Future Prospects

One of the primary aims of the research program described in this review has
been to formulate adequate two-reactant reactivity concepts, and the underlying
coordinate systems, which can be used to diagnose reactivity and selectivity
trends in systems of very large donor/acceptor reactants, e.g., chemisorption
systems. The CSA approach [52], which provides the basis for the present work,
is both relevant and attractive from the chemist's point of view, since many
branches of chemistry—the theory of chemical reactivity in particular—consider responses of chemical species to perturbations of the external potential and

D

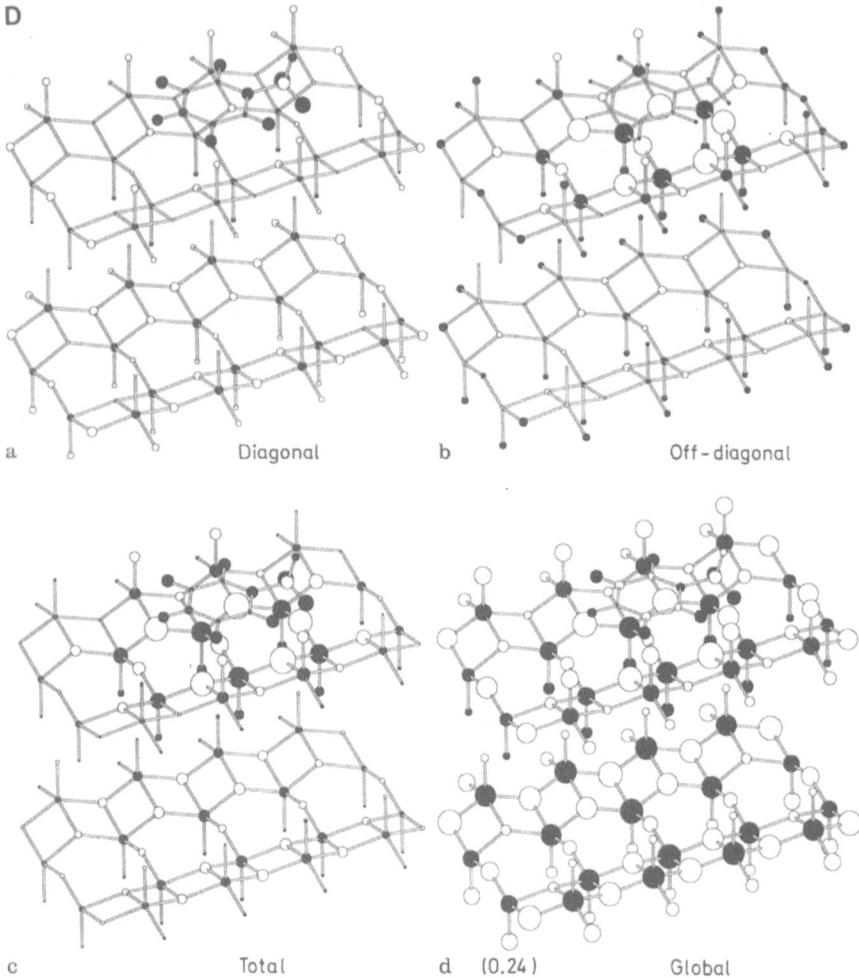

a   Diagonal   b   Off - diagonal

c   Total   d   (0.24)   Global

Fig. 34. (Conclusion). The isoelectronic, *in situ* FF indices (**Panels a–c**) and the global FF indices (**Panel d**) for the four adsorption arrangements of Fig. 33. The **Parts A, B, C, D** correspond to the structures c, d, e, f of Fig. 33, respectively. For reasons of clarity both layers have been mutually shifted by an arbitrary translation. The same scaling factor has been used in **Panels a–c**, while in **Panel d** the reported number is the factor required to bring this diagram to the same scale factor value as that used in the remaining panels. **Panels a, b** and **c** show (see caption to Fig. 9) the diagonal ($f^{A,A} - f^{B,B}$), off-diagonal ($f^{A,B} - f^{B,A}$), and total *in situ* FF indices for the assumed B (toluene) → A (cluster) electron transfer. The diagram of **Panel d** represents the corresponding global FF distribution, characterizing the external CT to the whole chemisorption system from an external reservoir

electron populations. Clearly, a two-reactant approach is vital, since the charge-response reactivity criteria and the interaction energy must involve measures of both the generalized polarizabilities of the perturbed reactant and the stimuli due to the presence of the other reactant. The CSA treats molecular systems as an interconnected ensemble of structural or functional units, with hypothetical flexibility and to open or close any of its constituent parts, to one another and/or relative to an external reservoir. It adopts a thermodynamic-like description [53] of the global and regional (constrained) equilibrium charge distributions, based on the chemical potential (electronegativity) equalization principle [3, 43, 54]. Such a perspective is indeed very close to that adopted in intuitive chemistry [54].

Each reactivity problem requires an appropriate level of resolution (reactants, functional groups or other important structural units, local sites, e.g., AIM, etc.). The canonical AIM-division can be expected to be sufficient for most chemical problems, although, for some subtle bonding effects, local [3, 21, 22, 55, 56] or molecular orbital [57, 58, 81] resolutions may be needed. Both function space (populational analysis) [3, 7, 9, 10, 12, 16, 59–62] and physical space (topological approach) [63–65] partitioning of the electronic density can be used to extract the AIM quantities. The AIM-type modelling has an *a priori* character, while the topological approach of Bader represents an *a posteriori* extraction of the AIM information from exact molecular calculations. When considering bond dissociation, one has to take into account chemical potential discontinuity [65, 66]. The semi-empirical modelling of the AIM charge sensitivities have recently been replaced by accurate *ab initio* wavefunction and DFT calculations [3, 56, 64, 65, 67–73].

The CSA in AIM resolution can be applied as a procedure supplementary to the standard *ab initio* SCF CI or DFT calculations, which may provide the input AIM charges, but this would severely limit a range of its applications. The real strength of the method lies in its ability to diagnose rapidly the reactivity trends from rather crude, and limited information about the constituent atoms, including the hardness (electron repulsion) interaction between large reactants, that depend parametrically on the geometrical structure of the system. Thus, even approximate AIM charges, e.g., those from the EEM, parametrized to reproduce STO-3G quality Mulliken populations, or from semi-empirical SCF MO calculations on small models, may be preferred, in order to explore, the reactivity possibilities between very large reactants quickly.

Each physical phenomenon can be related to the system of coordinates, in which it can be described in the most compact and simple way. Finding such an optimum reference frame is very important for interpretative purposes, since it identifies the crucial elements of the mechanism, and separates them from background information, that is irrelevant to the process in question – at least in the first approximation. We have shown that in reactive systems the highest degree of concentration of reactivity information is obtained in the reactant collective populational coordinates, which reflect the inter-reactant charge couplings, e.g., the IRM. Each system of coordinates probes a different facet of molecular behavior. Some of them, e.g., MEC/REC, are closely related to intuitive chemical thinking, *viz.*, inductive effect, charge relaxation accompanying localized reduction/oxidation, effects of a hypothetical CT between selected subsystems, etc.

R.F. Nalewajski, J. Korchowiec and A. Michalak

Although most of the sensitivity criteria are closely related to the interaction energy, they are basically of the response, geometrical character, reflecting the reaction stimulus. i.e., the interaction between the two molecules. The mode resolved CSA may also be considered as supplementing the classical orbital interaction approach of Fukui [11].

The main purpose of this review was to present a variety of concepts and the growing potential of the CSA for probing chemical reactivity. The method basically starts with the information on isolated reactants, which it supplements with the hardness interaction, interpolated from the reactant data and their current mutual geometrical arrangement in the reactive system. Thus, the CSA provides a phenomenological treatment of molecular systems, which allows one to make a quick prediction of their behavior in a changed environment, e.g., in the presence of the other molecule. This novel way of thinking about chemical reactivity was possible due to the DFT. In this outlook we have emphasized the two-reactant reactivity criteria. As we have demonstrated, the CSA enables one to formulate new geometric concepts for describing charge rearrangements between complex reactants [7–9, 74], including the alternative collective modes, which we have summarized in this review, and the intersecting-state-model for CT processes, defined in the AIM electron population space [6, 7, 25, 74]. This development is still in its experimental stage; most of the new concepts still await their eventual application to reactive/catalytic systems of real chemical interest. Also, the true potential behind the collective charge coordinate systems and the mapping relations remains to be explored.

Acknowledgments: The R.F.N. would like to thank Profs. W.J. Mortier and R.A. Schoonheydt and members of their research group at the Catholic University in Leuven, Drs. B.G. Baekelandt, H. Toufar and Mr. G.A.O. Jansens, for many stimulating discussions, particularly on the mapping relations. The authors would also like to thank Prof. K. Jug and members of his Laboratory at the University of Hannover, Drs. A.M. Köster and T. Bredow and Mr. H. Gerwens, for their help in generating the rutile cluster data, as well as Prof. M. Witko and Dr. R. Tokarz from the Institute of Catalysis and Surface Chemistry of the Polish Academy of Sciences, for providing us with structures and charges of the vanadium oxide clusters and chemisorption systems. The authors gratefully acknowledge research support from the State Committee for Scientific Research in Poland, the Commission of European Communities (COST D3 and D5 Actions), and the Exxon Chemical International Inc.

Note added in proof

We have used in Section 6.2 the *in situ* FF diagrams of Fig. 34 (A-c) – (D-c), calculated for the *assumed* toluene → cluster electron transfer, to interpret the toluene activation/surface reconstruction trends. Clearly, the CSA predicted direction of this inter-reactant CT (Eq. (177)) may be different from that assumed in these diagrams. Moreover, due to the changing interaction with an increase in the inter-reactant separation, which influences both μ_{CT}^+ and η_{CT}, such a geometrical change may also result in a change in the CT direction. A further interpretative limitation of the *in situ* FF diagrams is that one has to multiply them by N_{CT} to obtain the *absolute* AIM population changes due to the CT-component: $\{dN_X^*(N_{CT}), X = A, B\}$. Therefore, a full CSA interpretation of the CT reactivity trends requires a determination of the actual amount and direction of CT; it should be emphasized that the very identification of the acidic (A) and basic (B) reactants in $M = (A|B)$ also depends upon the actual direction of CT. This missing information can be calculated within CSA without difficulty (Eqs. (128) and (177)), provided the changes in the external potentials in atomic positions, due to the presence of the other reactant are known, since they uniquely specify the displacements in the AIM chemical potentials in the non-polarized reactants ("frozen" charge distributions of isolated reactants) [79, 82]. In the AIM description these perturbations can be realistically predicted from the point-charge approximation using the known AIM net charges in the isolated reactants: $dv_A = \{dv_a \cong -\sum_b^B q_b/|\vec{R}_a - \vec{R}_b|\}$ and $dv_B = \{dv_b \cong -\sum_b^A q_a/|\vec{R}_b - \vec{R}_a|\}$.

A knowledge of these realistic perturbing potentials also allows one to determine the remaining missing component of the charge reorganization in reactive systems, the direct polarization, *P-component*, $\{dN_X^+(dv), X = A, B\}$, defined by Eq. (180). These P-patterns allow one to qualitatively predict changes in the electronic structure of both reactants at the polarization stage, before the inter-reactant CT, and the adsorbate activation/surface reconstruction trends they imply [79, 82]. In addition one obtains the corresponding estimates of the ES, P and CT contributions to the interaction energy E_{CT} (see Eqs. (176), (177)), $E_{ES} = \sum_a^A \sum_b^B q_a q_b/|\vec{R}_b - \vec{R}_a|$, and $E_p = \frac{1}{2}dv\beta dv^\dagger$. These approximate energy estimates allow one to establish a rough energetical hierarchy among the chemically interesting chemisorption structures at various distances and reaction stages [79, 82].

Obviously, the full treatment of charge reorganizations in reactive systems calls for the overall (P + CT) – diagrams of the AIM charge displacements: $dN(dv, N_{CT}) = \{dN_X^+(dv) + dN_X^*(N_{CT}), X = A, B\}$. Only such diagrams can reveal which component (P or CT) dominates the overall (P + CT) pattern, thus playing the crucial role in the chemical reactivity. Using the P-patterns and the absolute CT-diagrams obtained from the point charge approximation to dv, one can address this problm as well, practically without any additional computational effort beyond that required to determine the *in situ* FF plots [79, 82].

In Fig. 35 we have compared the absolute charge displacement diagrams at all three levels of description, for the four chemisorption structures of Figs. 33 and 34. It follows from the CT patterns that at larger adsorbate-substrate separations (a, c, d) the direction of the inter-reactant CT is indeed the same as that assumed in Fig. 34, i.e., from toluene (B) to the surface cluster (A). However, at the closer approach (b) it is reversed, so that the cluster acts as a base and toluene becomes an acidic reactant. When the distance in panels (a, c, d) is decreased by 1 Å, to become comparable to that in panel (b), a similar reverse CT is observed in the parallel structures (c, d), but the CT direction stays

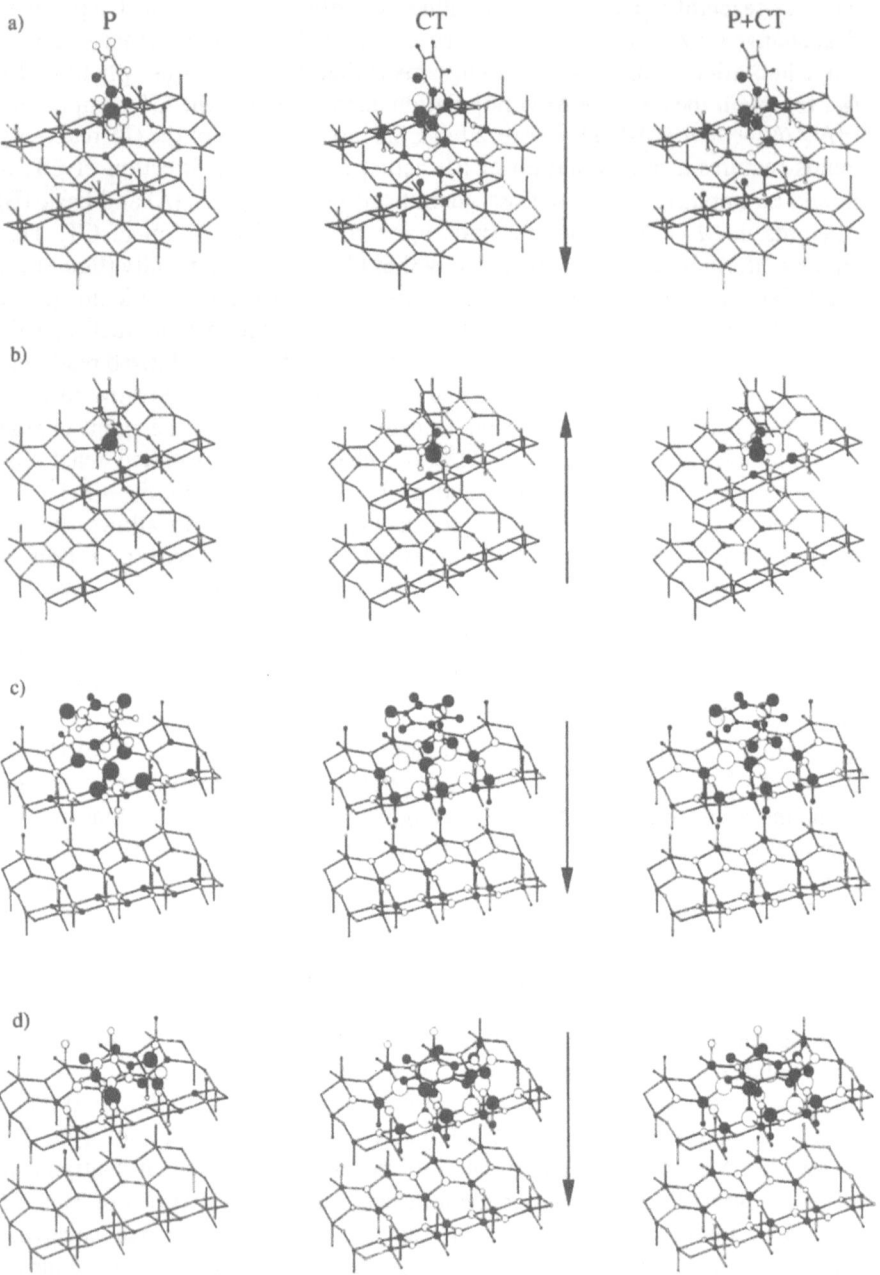

Fig. 35. The polarizational (P), charge transfer (CT), and overall (P + CT) charge reorganization patterns for the four chemisorption arrangements of Figs. 33 and 34, obtained from the CSA calculations using the point charge approximation to dν. The arrow in the CT components indices the CSA predicted direction of the charge flow. In each row different scale factors are applied in the P, CT, and (P + CT) panels.

unchanged in the complex (a); similarly, when the inter-reactant separation in diagram (b) is increased by 1 Å [to become comparable to that in complexes (a, c, d)] the CSA predicted CT direction, toluene → cluster, is identical to those observed in the remaining chemisorption complexes at these larger distances.

It follows from the comparison of Fig. 35 that the CT-component does indeed dominate the overall patterns in the *early approach structures* (a, c, d), while in the *close approach complex* (b) an increased role of the P-component can be detected. In the parallel adsorption arrangements the polarization patterns are seen to predict a stronger differentiation of the carbon charges in the ring and the opposite methyl polarization trends, in comparison to those exhibited by the CT-component. Large differences are also observed in the surface reconstruction due to the P and CT induced charge displacements, respectively, particularly in the complexes (b, c). In the perpendicular structure of Fig. 35a the methyl polarizations at the P and CT stages, respectively, are qualitatively different with the latter determining the overall trends. A similar difference is observed in the P and CT charge displacements of the $O_{(2)}$ adsorption site in Fig. 35b.

The total (ES + P + CT) interaction energies for the four structures of Fig. 35 indicate a strong preference towards the first (a) perpendicular adsorption on $O_{(2)}$ at both distances; among the parallel structures the first (c), central adsorption of the toluene ring above the surface vanadium site (c), is also energetically more favourable at both inter-reactant separations. The CSA energy estimates show that at the close approach the perpendicular structure (a) is the most stable one; at larger separations the parallel arrangement (c) is preferred. Clearly this energetical hierarchy changes when separate contributions to the inreaction energy are compared.

8 References

1. Hohenberg P, Kohn W (1964) Phys Rev B136: 864
2. Kohn W, Sham LJ (1965) Phys Rev A140: 1133
3. Parr RG, Yang W (1989) Density-Functional Theory of Atoms and Molecules. Oxford University Press, New York
4. Dreizler RM, Gross EKU (1990) Density Functional Theory: An Approach to the Quantum Many-Body Problem. Springer-Verlag, Heidelberg
5. Gross EKU, Draizler RM (eds) (1995) Proceedings of a NATO ASI on Density Functional Theory, Il Ciocco, August 16–27, 1993. Plenum Press, New York
6. Sen KD (ed) (1993) Structure and Bonding: Chemical Hardness. Springer-Verlag, Heidelberg
7. Nalewajski RF (1995) in: Gross EKU, Dreizler RM (eds) Proceedings of a NATO ASI on Density Functional Theory, Il Ciocco, August 16–27, 1993. Plenum Press, New York p 339–389
8. Nalewajski RF (1995) Int J Quantum Chem 56: 453
9. Nalewajski RF (1993) in: Sen KD (ed) Structure and Bonding, Volume 80: Chemical Hardness. Springer-Verlag, Heidelberg p 115–186; see also: Nalewajski RF, Korchowiec J (1989) J Mol Catal 54: 324
10. Baekelandt BG, Mortier WJ, Schoonheydt RA (1993) in: Sen KD (ed) Structure and Bonding, Volume 80: Chemical Hardness. Springer-Verlag, Heidelberg p 187–227

11. Fukui K (1973) Theory of Orientation and Stereoselection. Springer-Verlag, Heidelberg; (1982) Science 218: 747
12. Parr RG, Yang W (1984) J Am Chem Soc 106: 4049, Yang W, Parr RG (1985) Proc Natl Acad Sci USA 82: 6723
13. Falicov LM, Somorjai GA (1985) Proc Natl Acad Sci USA 82: 2207
14. Nalewajski RF, Korchowiec J, Zhou Z (1988) Int J Quantum Chem Symp 22: 349
15. Nalewajski RF (1990) Acta Phys Polon A77: 817
16. Antonova T, Neshev N, Proinov EI, Nalewajski RF (1991) Acta Phys Polon A79: 805
17. Pariser R (1953) J Chem Phys 21: 568
18. Ohno K (1967) Adv Quantum Chem 3: 239; (1968) Theoret Chim Acta 10: 111
19. Nishimoto K, Mataga N (1957) Z Physik Chem 12: 335; Mataga N, Nishimoto K (1957) Z Physik Chem 13: 140
20. Korchowiec J, Gerwens H, Jug K (1994) Chem Phys Lett 222: 58
21. Chattaraj PK, Parr RG (1993) in: Sen KD (ed) Structure and Bonding Volume 80: Chemical Hardness. Springer-Verlag, Heidelberg p 11–25; Parr RG, Chattaraj PJ (1991) J Am Chem Soc 113: 1854
22. Berkowitz M, Parr RG (1988) J Chem Phys 88: 2554
23. Baekelandt BG, Jansens GOA, Toufar H, Mortier WJ, Schoonheydt RA, and Nalewajski RF, (1995) J Phys Chem 99: 9784
24. Nalewajski RF (1991) Int J Quantum Chem 40: 265; (1992) Int J Quantum Chem 43: 443; see also Ref. 26
25. Nalewajski RF (1994) Int J Quantum Chem 49: 675
26. Nalewajski RF, Korchowiec J (1993) J Mol Catal 82: 383
27. Nalewajski RF, Michalak A (1995) Int J Quantum Chem 56: 603
28. Nalewajski RF, Korchowiec J (1995) Computers Chem 19: 217
29. Nalewajski RF, Koniński M (1987) Z Naturforsch 42a: 451
30. Decius JC (1963) J Chem Phys 38: 241; Jones LH, Ryan RR (1970) J Chem Phys 52: 2003
31. Swanson BI (1976) J Am Chem Soc 98: 3067; Swanson BI, Satija SK (1977) J Am Chem Soc 99: 987
32. Dewar MJS, Thiel W (1977) J Am Chem Soc 99: 4899
33. Gutmann V (1978) The Donor-Acceptor Approach to Molecular Interactions. Plenum Press, New York
34. RG Pearson (1976) Symmetry Rules for Chemical Reactions: Orbital Topology and Elementary Processes. Wiley, New York and references therein
35. Bader RFW (1960) Mol Phys 3: 137; Bader RFW, Bandrauk AD (1968) J Chem Phys 49: 1666
36. Nakatsuji H (1974) J Am Chem Soc 96: 24, 30
37. Janssens GAO, Baekelandt BG, Toufar H, Mortier WJ, Schoonheydt RA (1995) Int J Quantum Chem 56: 317
38. Parr RG, Borkman RF (1968) J Chem Phys 49: 1055
39. Nalewajski RF (1989) in: J. Popielawski (ed) Proceedings of the International Symposium on the Dynamic of Systems with Chemical Reaction, Świdno, June 6–10, 1988. World Scientific, Singapore p 325
40. Nalewajski RF, Korchowiec J (1989) Acta Phys Polon A76: 747; Korchowiec J, Nalewajski RF (1992) Int J Quantum Chem 44: 1027
41. Gázquez JL (1993) in: Sen KD (ed) Structure and Bonding Volume 80: Chemical Hardness. Springer-Verlag, Heidelberg p 27–44
42. Morokuma K (1977) Acc Chem Res 10: 294
43. Parr RG, Donnelly RA, Levy M, Palke WE (1978) J Chem Phys 68: 801; Donnelly RA, Parr RG (1978) J Chem Phys 69: 4431
44. Witko M, Tokarz R, Haber J (1991) J Mol Catal 66: 205
45. Nalewajski RF, Korchowiec J, Michalak A (1994) Gadre S (ed) Proc Indian Acad Sci (Chemical Sci) 106: 353
46. Gołębiewski A (1961) Trans Faraday Soc 57: 1849
47. Nalewajski RF (1993) J Mol Catal 82: 371
48. Nalewajski RF, Köster AM, Bredow T, Jug K (1993) J Mol Catal 82: 407
49. Kurtz RL, Stockbauer R, Madey TE, Román E, de Segovia JL (1989) Surf Sci 218: 178
50. Lo WJ, Chung YW, Somorjai GA (1978) Surf Sci 71: 199
51. Nanda DN, Jug K (1980) Theoret Chim Acta 57: 95; Jug K, Iffert R, Schulz J (1987) Int J Quant Chem 32: 265; Li J, Jug K (1992) J Comput Chem 13: 85

52. Nalewajski RF, Korchowiec J (1996) Charge Sensitivity Approach to Molecular Structure and Chemical Reactivity. World Scientific, Singapore; prepared for publication
53. Callen HB (1960) Thermodynamics: An Introduction to the Physical Theories of Equilibirum Thermostatics and Irreversible Thermodynamics. Wiley, New York
54. Sanderson RT (1952) J Am Chem Soc 74: 272; (1976) Chemical Bonding and Bond Energy. Academic Press, New York
55. Nalewajski RF, Parr RG (1982) J Chem Phys 77: 399; Nalewajski RF (1983) J Chem Phys 78: 6112; Nalewajski RF (1988) Z Naturforsch 43a: 65; Nalewajski RF, Capitani JF (1982) J Chem Phys 77: 2514; Nalewajski RF (1985) J Phys Chem 89: 2831
56. Cohen MH, Ganduglia-Pirovano MV, Kurdnovský J (1994) J Chem Phys 101: 8988; (1994) Phys Rev Lett 72: 3222; J Chem Phys, In press; see also the next paper of this volume
57. Nalewajski RF, Mrozek J (1992) Int J Quantum Chem 43: 353
58. Nalewajski RF (1992) Int J Quantum Chem 44: 67
59. Nalewajski RF, Koniński M (1984) J Phys Chem 88: 6234; (1988) Acta Phys Polon A74: 255; Nalewajski RF (1990) Acta Phys Polon A77: 817
60. Li L, Parr RG (1986) J Chem Phys 84: 1704; Rychlewski J, Parr RG (1986) J Chem Phys 84: 1696
61. Mortier WJ, Ghosh SK, Shankar S (1986) J Am Chem Soc 108: 4315; Mortier WJ (1987) in: Sen KD, Jorgensen CK (eds) Structure and Bonding Volume 66: Electronegativity. Springer Verlag, Heidelberg p 125–143; Mortier WJ, van Genechten K, Gasteiger J (1985) J Am Chem Soc 107: 829
62. Komorowski L, Lipiński J (1991) Chem Phys 157: 45; Komorowski L (1993) in: Sen KD (ed) Structure and Bonding, Volume 80: Chemical Hardness. Springer-Verlag, Heidelberg p 45–70
63. Bader RFW (1990) Atoms in Molecules: A Quantum Theory. Clarendon Press, Oxford
64. Ciosłowski J, Stefanov BB (1993) J Chem Phys 99: 5151
65. Ciosłowski J, Mixon ST (1993) J Am Chem Soc 115: 1084; Cioslowski J (1994) Int J Quantum Chem 49: 463; Ciosłowski J, Martinov M (1994) J Chem Phys 101: 366
66. Perdew JP, Parr RG, Levy M, Balduz JL (1982) Phys Rev Lett 49: 1691; see also: Perdew JP (1985) in: Dreizler RM, da Providência J (eds) Proceedings of a NATO ASI on Density Functional Methods in Physics, Alcabideche, September 5–16, 1983. Plenum Press, New York p 265–308
67. Lee C, Yang W, Parr RG (1988) J Mol Struct (Theochem) 163: 305
68. Gázquez JL, Vela A, Galván M (1987) in: Sen KD, Jorgensen CK (eds) Structure and Bonding Volume 66: Electronegativity. Springer Verlag, Heidelberg p 79
69. De Proft F, Langenaeker W, Geerlings P (1993) J Phys Chem 97: 1826; Langenaeker W, Demel K, Geerlings P (1991) J Mol Struct (Theochem) 234: 329
70. Bartolotti LJ, Gadre SR, Parr RG (1980) J Am Chem Soc 102: 2945; Robles J, Bartolotti LJ (1984) J Am Chem Soc 106: 3723
71. Sen KD (1993) in: Sen KD (ed) Structure and Bonding, Volume 80: Chemical Hardness. Springer-Verlag, Heidelberg p 87–100
72. March NH (1993) in: Sen KD (ed) Structure and Bonding, Volume 80: Chemical Hardness. Springer-Verlag, Heidelberg p 71–86
73. Alonso JA, Balbás LC (1993) in: Sen K (ed) Structure and Bonding, Volume 80: Chemical Hardness. Springer-Verlag, Heidelberg pp 229–258
74. Nalewajski RF (1992) Int J Quantum Chem 42: 243
75. Nalewajski RF (1996) Int J Quantum Chem, in press
76. Nalewajski RF, Korchowiec J, Tokarz R, Brocławik E, Witko M (1992) J Mol Catal 77: 165
77. Witko M, Haber J, Tokarz R (1993) J Mol Catal 82: 457
78. Michalak A, Hermann K, unpublished results
79. Nalewajski RF, Korchowiec J (1995) J Mol Catal, in press
80. Nalewajski RF, Mrozek J, Michalak A (1996) Int J Quantum Chem, in press
81. Nalewajski RF, unpublished results
82. Nalewajski RF, Michalak A (1996) J Phys Chem, submitted.

Strengthening the Foundations of Chemical Reactivity Theory

Morrel H. Cohen

Corporate Research Laboratories, Exxon Research and Engineering Company, Route 22 East, Annandale, New Jersey 08801, USA

Table of Contents

Topics in Current Chemistry, Vol. 183
© Springer-Verlag Berlin Heidelberg 1996

Chemical reactivity theory, in its evolution from its earliest empirical beginnings to its current level of sophistication has had an immense impact on chemistry. In work of Parr and collaborators in the 1980s, the theory was grounded in a rigorous formulation of many-body theory and density functional theory, and its richness and promise were made manifest in contemporaneous work by Nalewajski and collaborators. Nevertheless, many fundamental issues remained unresolved, and the need to strengthen the foundations of reactivity theory and broaden its applicability remained. In the present chapter, the early work of Parr, Nalewajski and their coworkers is reviewed and the unresolved issues defined and enumerated. Recent work by the author and his collaborators addressing the resolution of these issues is reviewed, including both published and previously unpublished results. The principal advances discussed include explicit Kohn-Sham expressions for reactivities, the distinction between localized and extended chemical systems and the associated limitation of the ensemble formulation of density-functional theory, the distinction between isoelectronic and electron-transfer reactivities, the need for both static and dynamic reactivities, the need for nonlocal reactivity kernels, the placement of nuclear and electronic reactivities on the same footing, the identification of chemical stimuli in addition to chemical responses, and the establishment of a relationship between the total energy and the reactivities and, ultimately, the reaction pathway. While many elements of the theory have been completed, the proper definition of electron-transfer reactivities for localized systems and the relation between total energy and reactivities for overlapping reactants have not yet been found. How this might be done is briefly discussed, and it is suggested that successfully accomplishing it would complete the strengthening of the foundations of reactivity theory attempted by the author and collaborators.

1 Introduction

The formulation of concepts with which to order the vast phenomenology of chemical reactions and, ultimately, to predict their course has been a focal activity of theoretical chemistry since its inception. These concepts have been transformed over the years from empirical notions into what has become a rich, multifaceted structure conveniently labeled chemical-reactivity theory. In the course of its development, reactivity theory has had an immense impact on chemistry. Given its importance and the advanced state to which it has been brought, as reviewed, e.g. by Nalewajski, Korchowiec and Michalak, the preceding chapter in this volume [1], it seemed worthwhile to essay a fully rigorous development of the structure of reactivity theory in an effort to strengthen its foundations. As will be seen, that task is not yet complete, but in the present chapter we review our progress to date, including both published [2, 3] and as yet unpublished material by the author and his collaborators.

Of all the many important contributions to reactivity theory, two stand out as particularly relevant to our task. The first is Fukui's introduction in 1952 of the concept of frontier orbitals within an independent-electron framework [4–7] (of which the Hartree-Fock approximation was of course the most accurate realization) which gave a clear, quantum-mechanical basis to notions that had previously been semiempirical. The second, from Parr and collaborators [8–15], showed in the 1980s how Fukui's and other, closely related ideas could be given a rigorous foundation in density-functional theory [8, 16–18].

Of the central concepts emerging from the work of Parr and collaborators and of Nalewajski and collaborators [1, 19–24], reactivity indices such as the Fukui function [8–12], the local and global softnesses [8–12], the corresponding hardnesses [13–14], and the hardness and softness kernels [15, 19–21] provide quantitative measures of the reactive proclivities of isolated, unperturbed systems. Expressing the interaction energy of two subsystems in terms of these indices then shows how their individual proclivities mesh. While many earlier workers dealt with two-reactant systems (cf. references in [1]), our goal here is to do so again but with complete rigor, subject to the limitation that the interaction energy is expressed entirely in terms of such properties of the isolated reactants as the reactivity indices, a defining feature of reactivity theory. This limitation restricts consideration of the interaction energy to a low order of perturbation theory, that is, to weakly interacting systems. The common goal of quantitative theories of chemical reactions is to calculate reaction rates from a knowledge of energy surfaces in configuration space (see e.g. [25]). Because of the above limitation, chemical reactivity theory can give quantitative information only about the early stages of chemical reactions, i.e., about the portal or gateway to the reaction pathway, unless the entire pathway remains in the weak interaction regime. Certain electron-transfer reactions provide examples of this most favorable case, and it would provide a rigorous basis for the atoms-in-molecules and molecular fragment pictures.

Nevertheless, a rigorous treatment of the early stages of a reaction could be most valuable in the general case. Even for a system computationally simple enough for its energy surface to be calculated in sufficient detail for a reaction-rate computation to be feasible, interpretation of the results can be greatly facilitated through the use of reactivity theory. Moreover, reactivity theory can be used to generate predictions testable by computation and experiment. There are, of course, systems too complex computationally either for a direct attack on their energy surfaces or for accurate calculation of their reactivity indices. Even so, there are also systems of an intermediate level of computational complexity for which the system as a whole is too complex for direct attack while computation of the reactivity indices of the individual reactants is still feasible. For such systems, reactivity theory can be of great usefulness.

Heterogeneous catalysis provides many important examples of those intermediate cases [26]. Accurate numerical study of the interaction of very simple molecules with the clean, atomically flat surfaces of pure metals has only very recently become feasible [27–34]. For example, the full six-dimensional energy surfaces of H_2 on clean [33] and sulfur-poisoned [34] Pd(100) surfaces have been calculated. The energy surfaces which result from these computations have been used to compute the temperature-dependent reactive sticking coefficient of H_2 on Cu(111) [35] and Pd(100) surfaces [36]. These calculations push the boundary of what is now feasible. Any significant complication, e.g. an alloy surface or a somewhat larger molecule, completely inhibits such a direct attack from first principles at this time. On the other hand, the reactivity indices of an alloy surface could now be calculated and so could those of a considerably larger molecule. Good use could be made at present, then, of a theory which defines reactivity indices and relates them to reactant interaction energies in a fully rigorous manner in this and many other instances. Chemical reactivity theory should therefore be developed to the point where it can thus be used to finesse chemical complexity. In the following, we review efforts by the author and collaborators to strengthen the foundations of chemical reactivity theory and broaden its applicability so that it can be so used.

In Sect. 2 we introduce Fukui's frontier-orbital notion [4–7] and briefly review the grounding of this notion in density-functional theory by Parr and collaborators [8–12] through their introduction of the Fukui function and the local and global softnesses. During the course of our subsequent presentation, we point out a series of issues fundamental to this grounding of chemical-reactivity theory in density-functional theory [8, 16–18]. Sections 3 through 8 are devoted to elucidating these issues, and reviewing their resolution to date. In Sect. 3, we give explicit expressions for the reactivity indices defined in Sect. 2 within the framework of the theory of Kohn-Sham [17]. In Sect. 4, we point out that a clear distinction must be made between localized systems with discrete electronic spectra and extended systems with quasicontinuous or continuous spectra. We show that the reactivity indices of Sect. 2 can actually be defined only for extended systems. We then review in Sect. 5 analogous concepts of nuclear reactivity valid for extended systems, nuclear rearrangement being the defining basis of a chemical reaction. We point out in Sect. 6 that the reactivity indices of Sect. 2 are local responses to global perturbations, whereas chemical reactions are nonlocal responses to local perturbations. We then review the reactivity kernels introduced by Nalewajski [19–21] and by Berkowitz and Parr [15] which resolve this issue. We give Kohn-Sham expressions for the softness kernels and introduce the distinction between isoelectronic and charge-transfer softness kernels. As emerges from the work of Nalewajski and collaborators [22, 23], these reactivity kernels also resolve the ambiguity of the frontier-orbital concept for extended systems and display clearly the complexity of chemical responses; we discuss this issue in Sect. 7. Finally, in Sect. 8, we show how analysis of the total energy of interaction of two disjoint systems through the second order of perturbation theory permits the identification of dynamic generalizations of the static isoelectronic chemical reactivities thus far introduced and of chemical stimuli as well. The status of the fundamental issues raised in Sects. 3 through 8 is summarized in Sect. 9, where it is pointed out that the most important task remaining is the generalization of the interaction-energy analysis of Sect. 8 to overlapping systems in order to establish charge-transfer reactivities analogous to the isoelectronic reactivities of Sect. 8 as well as establish the corresponding stimuli.

2 The Fukui Function and the Local and Global Softnesses

Fukui [4–7] introduced as quantitative measures of reactivity the frontier-orbital densities,

$$\rho_F^+ (\mathbf{r}) = \sum_s |\phi_{N+1}(\mathbf{x})|^2 , \tag{1a}$$

$$\rho_F^- (\mathbf{r}) = \sum_s |\phi_N(\mathbf{x})|^2 , \tag{1b}$$

of an N-electron system treated in an independent-electron approximation which could be semi-empirical or as sophisticated as Hartree-Fock. The electron coordinate x has a space component $\mathbf{r}$ and a spin component s. ϕ_{N+1} is the lowest unoccupied molecular orbital or LUMO, and ϕ_N is the highest occupied molecular orbital or HOMO. Thus $\rho_F^+(\mathbf{r})$ is presumed to measure the local reactivity to a nucleophilic reagent or donor, and $\rho_F^-(\mathbf{r})$ is presumed to measure the local reactivity to an electrophilic reagent or acceptor.

Parr and collaborators [8–12] showed how Fukui's frontier-orbital concept could be grounded in a rigorous many-electron theory, density-functional theory (DFT) [8, 16–18]. They used the ensemble formulation of DFT to introduce the expectation value $\mathcal{N}$ of the total electron number as a continuous variable. They then defined the Fukui functions

$$f^\pm (\mathbf{r}) = \left[\frac{\partial \rho(\mathbf{r})}{\partial \mathcal{N}} \right]_v^\pm , \quad \mathcal{N} = N_0 \pm 0^+, \tag{2}$$

as the reactivity indices corresponding to Fukui's frontier-orbital densities. The subscript v specifies that the differentiation is to be carried out at constant nuclear electrostatic potential $v(\mathbf{r})$ acting on an electron at $\mathbf{r}$ plus any other external potential. The superscript $+ (-)$ indicates that the derivative with respect to $\mathcal{N}$ is taken at $\mathcal{N}$ just above (below) an integer value N_0. The discontinuity between $f^+(\mathbf{r})$ and $f^-(\mathbf{r})$ arises from the discontinuities at integral values of $\mathcal{N}$ in the $T = 0$ grand canonical ensemble underlying the DFT used. Proceeding next via the Kohn-Sham (KS) reformulation of DFT [17], Parr and colleagues [8–12] expressed the Fukui functions in terms of the exact Kohn-Sham orbitals $\phi_i(\mathbf{x})$ [12],

$$f^\pm (\mathbf{r}) = \rho_F^\pm (\mathbf{r}) + \sum_{i=1,s}^{N_0} \left[\frac{\partial \psi_i(\mathbf{x})^2}{\partial \mathcal{N}} \right]_v^\pm . \tag{3}$$

The frontier-orbital density $\rho_F^\pm(\mathbf{r})$ in Eq. (3) is now defined in terms of the exact KS LUMO and HOMO instead of the approximate orbitals $\phi_i(x)$ in Eq. (1),

$$\rho_F^+(\mathbf{r}) = \sum_s |\psi_{N+1}(x)|^2 , \tag{4a}$$

$$\rho_F^-(\mathbf{r}) = \sum_s |\psi_N(x)|^2 . \tag{4b}$$

We see from Eq. (3) that in addition to its use of approximate orbitals, Fukui's frontier density is a frozen-orbital approximation to the Fukui function, as indicated by the second term on the right-hand side of Eq. (3) [32].

The Fukui function is normalized to unity,

$$\int d\mathbf{r} f^\pm(\mathbf{r}) = 1 . \tag{5}$$

Thus it describes only the local geometry of a nucleophilic $(-)$ or electrophilic $(+)$ reactive response; it does not describe the local intensity of response. The local softness, $s^\pm(\mathbf{r})$, defined by Parr and collaborators [8–12, 33] as

$$s^\pm(\mathbf{r}) = \left[\frac{\partial \rho(\mathbf{r})}{\partial \mu}\right]_v^\pm \tag{6}$$

does describe the intensity of response. Here, μ is the chemical potential or Fermi level of the electrons. The global softness is then defined as

$$S^\pm = \int d\mathbf{r} s^\pm(\mathbf{r}) . \tag{7}$$

From Eqs. (6) and (7) it follows that

$$S^\pm = \left[\frac{\partial \mathcal{N}}{\partial \mu}\right]_v^\pm \tag{8}$$

since

$$\mathcal{N} = \int d\mathbf{r} \rho(\mathbf{r}). \tag{9}$$

The Fukui function is thus a normalized local softness,

$$f^\pm(\mathbf{r}) = s^\pm(\mathbf{r})/S^\pm . \tag{10}$$

In addition to the local and global softnesses, Ghosh, Berkowitz and Parr also defined local and global hardnesses [13–15], $\eta^\pm(\mathbf{r})$ and $\eta^\pm$ [39]

$$2\eta^\pm(\mathbf{r}) = \left[\frac{\delta \mu}{\delta \rho(\mathbf{r})}\right]_v^\pm, \tag{11}$$

$$\eta^\pm = \int d\mathbf{r} \eta^\pm(\mathbf{r}) f^\pm(\mathbf{r}) . \tag{12}$$

The local hardness and softness are reciprocal quantities in the following sense [13–15],

$$2\int d\mathbf{r} \eta^\pm(\mathbf{r}) s^\pm(\mathbf{r}) = 1 , \tag{13}$$

as are the global quantities, since Eqs. (2), (12), and (8) imply that

$$\eta = 1/2S . \tag{14}$$

The local and global hardnesses and more general hardnesses to be introduced later are very important quantities. Nevertheless, we do not emphasize them in the present chapter because, for the most part, the issues that arise at the foundations of the theory can be dealt with by considering the corresponding softnesses.

3 Kohn-Sham Expressions for $f^{\pm}(r)$, $s^{\pm}(r)$, and S

The first issue to be dealt with is most readily framed as a question. Can one find explicit and exact expressions for the reactivity indices $f^{\pm}(r)$, $s^{\pm}(r)$, and $S^{\pm}$ in terms of quantities calculable within the Kohn-Sham (KS) framework? This question was answered in [2] in the affirmative. The analysis starts from the standard expression for the electron density in terms of the Kohn-Sham orbitals for an N_0-electron system

$$\rho(r) = \sum_{i=1,s}^{N_0} |\psi_i(x)|^2 . \tag{15}$$

This is most conveniently generalized to the ensemble formulation of Kohn-Sham theory [8, 18] by introducing the Kohn-Sham Green's function

$$G(x, x'; E^-) = \langle x|[E^- - H_{KS}]^{-1}|x'\rangle , \tag{16a}$$

where

$$E^- = E - i\delta, \quad \delta \to 0^+ . \tag{16b}$$

The KS Hamiltonian H_{KS} is

$$H_{KS} = \frac{p^2}{2m} + v_{KS}(r) , \tag{17a}$$

and the Kohn-Sham potential is

$$v_{KS}(r) = v(r) + v_H(r) + v_{xc}(r) . \tag{17b}$$

In Eq. (17b), $v(r)$ is, as before, the nuclear electrostatic potential acting on an electron at r; $v_H(r)$ is the Hartree potential

$$v_H(r) = \int dr' \frac{e^2}{|r - r'|} \rho(r') ; \tag{17c}$$

and $v_{xc}(r)$ is the KS exchange and correlation potential. The exchange and correlation potential is derived from the energy function [16]

$$E[\rho] = F[\rho] + \int dr\, v(r)\, \rho(r) \tag{18a}$$

of the electrons, where $F[\rho]$ is a universal functional, the sum of the electron kinetic-energy functional $T_e[\rho]$ and the electron-electron interaction functional $V_{ee}[\rho]$,

$$F[\rho] = T_e[\rho] + V_{ee}[\rho] . \tag{18b}$$

The explicit expression for $v_{xc}(\mathbf{r})$ is

$$v_{xc}(\mathbf{r}) = \frac{\delta F[\rho]}{\delta \rho(\mathbf{r})} - v_H(\mathbf{r}) - \frac{\delta T_S[\rho]}{\delta \rho(\mathbf{r})} , \tag{19}$$

where $T_S[\rho]$ is the kinetic energy functional of a system of independent electrons obeying the exclusion principle with a total electron density constrained to be $\rho(\mathbf{r})$ [8, 17, 18].

The analysis in [2] results in the following KS expressions for the reactivity indices $f^{\pm}(\mathbf{r})$ and $s^{\pm}(\mathbf{r})$

$$f^{\pm}(\mathbf{r}) = \int d\mathbf{r}' K^{-1}(\mathbf{r}, \mathbf{r}') \rho_F^{\pm}(\mathbf{r}') , \tag{20}$$

$$s^{\pm}(\mathbf{r}) = \int d\mathbf{r}' K^{-1}(\mathbf{r}, \mathbf{r}') g(\mathbf{r}', \mu) . \tag{21}$$

In Eq. (20), $\rho_F^{\pm}(\mathbf{r})$ is the frontier density, Eq. (4). In Eq. (21), $g(\mathbf{r}, \mu)$ is the local density of Kohn-Sham states (DOS) at the chemical potential [40],

$$g(\mathbf{r}, \mu) = \sum_s \frac{1}{\pi} \operatorname{Im} G(\mathbf{x}, \mathbf{x}; \mu^-) . \tag{22}$$

The kernel $K^{-1}(\mathbf{r}, \mathbf{r}')$ entering both Eqs. (20) and (21) is the inverse of

$$K(\mathbf{r}, \mathbf{r}') = \delta(\mathbf{r}, \mathbf{r}') + \int d\mathbf{r}'' P(\mathbf{r}, \mathbf{r}'') W(\mathbf{r}'', \mathbf{r}') . \tag{23}$$

In Eq. (23), $P(\mathbf{r}, \mathbf{r}')$ is the static polarization propagator,

$$P(\mathbf{r}, \mathbf{r}') = - \left[\frac{\delta \rho(\mathbf{r})}{\delta v_{KS}(\mathbf{r}')} \right]_{\mu} = P(\mathbf{r}', \mathbf{r}) \tag{24a}$$

$$= - \sum_{s,s'} \int^{\mu} dE \frac{1}{\pi} \operatorname{Im} [G(\mathbf{x}, \mathbf{x}'; E^-) G(\mathbf{x}', \mathbf{x}; E^-)] . \tag{24b}$$

The local DOS is related to $P(\mathbf{r}, \mathbf{r}')$ through

$$g(\mathbf{r}, \mu) = \int d\mathbf{r}' P(\mathbf{r}, \mathbf{r}') . \tag{25}$$

The quantity $W(\mathbf{r}, \mathbf{r}')$ plays the role of an effective electron-electron interaction; it is defined as

$$W(\mathbf{r}, \mathbf{r}') = \frac{\delta v_{KS}(\mathbf{r})}{\delta \rho(\mathbf{r}')} , \tag{26a}$$

$$= \frac{e^2}{|\mathbf{r} - \mathbf{r}'|} + W_{xc}(\mathbf{r}, \mathbf{r}') \tag{26b}$$

and termed the Kohn-Sham interaction. The quantity

$$W_{xc}(\mathbf{r}, \mathbf{r}') = \frac{\delta v_{xc}(\mathbf{r})}{\delta \rho(\mathbf{r}')} \tag{26c}$$

is an exchange and correlation interaction, symmetric in $\mathbf{r}$ and $\mathbf{r}'$ because it is a second functional derivative [cf. Eq. (19)], and it is of short range. $W(\mathbf{r}, \mathbf{r}')$ is thus asymptotically equal to the bare Coulomb interaction [41].

The kernel $K(\mathbf{r}, \mathbf{r}')$ is the transpose of $\mathscr{K}(\mathbf{r}, \mathbf{r}')$, the Kohn-Sham potential-response function (KSPRF),

$$K(\mathbf{r}, \mathbf{r}') = \mathscr{K}(\mathbf{r}, \mathbf{r}')^T \tag{27a}$$

$$\mathscr{K}(\mathbf{r}, \mathbf{r}') = \left[\frac{\delta v(\mathbf{r})}{\delta v_{KS}(\mathbf{r}')} \right]_\mu \tag{27b}$$

$$= \delta(\mathbf{r}, \mathbf{r}) + \int d\mathbf{r}'' W(\mathbf{r}, \mathbf{r}'') \, P(\mathbf{r}'', \mathbf{r}') \tag{27c}$$

$$\sim \delta(\mathbf{r}, \mathbf{r}') + \int d\mathbf{r}'' \frac{e_2}{|\mathbf{r} - \mathbf{r}''|} P(\mathbf{r}'', \mathbf{r}'). \tag{27d}$$

Equation (27d) states that the kernel $\mathscr{K}(\mathbf{r}, \mathbf{r}')$ is asymptotically equal to the Hartree-Kohn-Sham static dielectric function. Thus the expression in Eq. (20) for the Fukui function is just a short-range linear mapping of the frontier density, and the expression in Eq. (21) for the local softness is the same mapping of the local DOS. It is the frontier-orbital density which drives the chemical response measured by the Fukui function, and the local DOS which drives that measured by the local softness.

4 The Distinction Between Continuous or Quasicontinuous and Discrete Spectra

All finite systems have discrete energy spectra. This simple fact has major consequences for reactivity theory as it has been developed thus far. We therefore reexamine here the argument which underlies the treatment of the total number of electrons in a localized system, e.g. a molecule, as a continuous variable $\mathscr{N}$ despite the fact that an individual system can hold only an integral number of electrons [42].

4.1 Fukui Function and Softnesses of Localized Systems in the $T = 0$ Grand Canonical Ensemble

Let the temperatures of interest be sufficiently low as to be negligible compared to the relevant ionization potentials and electron affinities. One can then use the zero-temperature $(T = 0)$ limit of the grand-canonical ensemble (GCE) to characterize a system with a continuous average number $\mathcal{N}$. The corresponding density matrix is

$$D_{\mathcal{N}} = (1 - \varepsilon)\, \Psi^0_{N_0}(\Psi^0_{N_0})^* + \varepsilon\Psi^0_{N_0+1}(\Psi^0_{N_0+1})^* ,\tag{28a}$$

$$\mathcal{N} = N_0 + \varepsilon,\tag{28b}$$

$$0 \leq \varepsilon \leq 1, \quad N_0 \leq \mathcal{N} \leq N_0 + 1 .\tag{28c}$$

N_0 can be any integer to which the system of electrons is bound. In Eq. (28a), $\Psi^0_{N_0}$ is the ground-state wave function of the N_0-particle system and $\Psi^0_{N_0+1}$ that of the $(N_0 + 1)$-particle system. Thus $D_{\mathcal{N}}$ changes discontinuously as $\mathcal{N}$ crosses an integer. It is a continuous, piecewise-linear operator-valued function of $\mathcal{N}$. Consequently, all expectation values of operators are also piecewise-linear functions of $\mathcal{N}$. Their first derivatives with respect to $\mathcal{N}$ are all discontinuous, piecewise-constant functions of $\mathcal{N}$.

The average total energy $E(\mathcal{N})$ is

$$E(\mathcal{N}) = (1 - \varepsilon)E^0_{N_0} + \varepsilon E^0_{N_0+1} \quad N_0 \leqslant \mathcal{N} \leqslant N_0 + 1 ,\tag{29}$$

where the E^0 are the ground-state energies of the N_0- or $(N_0 + 1)$-particle system. The chemical potential is, therefore,

$$\mu(\mathcal{N}) = \left[\frac{\partial E(\mathcal{N})}{\partial \mathcal{N}}\right]_v = \left[\frac{\partial E}{\partial \varepsilon}\right]_v = E^0_{N_0+1} - E^0_{N_0}$$

$$= -A_{N_0} = -I_{N_0+1}, \quad N_0 < \mathcal{N} < N_0 + 1;\tag{30a}$$

$$-I_{N_0} \leq \mu \leq -I_{N_0+1}, \quad \mathcal{N} = N_0 .\tag{30b}$$

Here, the symbol A represents an electron affinity and I an ionization energy with their subscripts indicating the relevant electron number. Equations (30a, b) show that $\mu(\mathcal{N})$ is a piecewise-constant function of $\mathcal{N}$, multivalued at $\mathcal{N}$ an integer. The ground-state energy $E^0_{N_0}$ is known but not proven to be convex in N_0 in the sense that its second difference is positive [8],

$$(E^0_{N_0+1} - 2E^0_{N_0} + E^0_{N_0-1}) > 0 ,\tag{31}$$

which implies that

$$I_{N_0} > I_{N_0+1} .\tag{32}$$

Thus $\mu(\mathcal{N})$ is an ascending staircase. The physical interpretation of $\mu(\mathcal{N})$ is quite simple. The ensemble of Eq. (28) describes an assembly consisting only of N_0-electron systems $(\varepsilon = 0)$ for $-I_{N_0} < \mu < -I_{N_0+1}$ and of a physical admixture of N_0- and $(N_0 + 1)$-electron systems when $\mu = -I_{N_0+1} (0 < \varepsilon < 1)$. $\mu(\mathcal{N})$ can be inverted to give $\mathcal{N}(\mu)$, a piecewise-constant function of μ multivalued at the ionization energies, also a staircase. This $\mathcal{N} - \mu$ staircase is illustrated in Fig. 1.

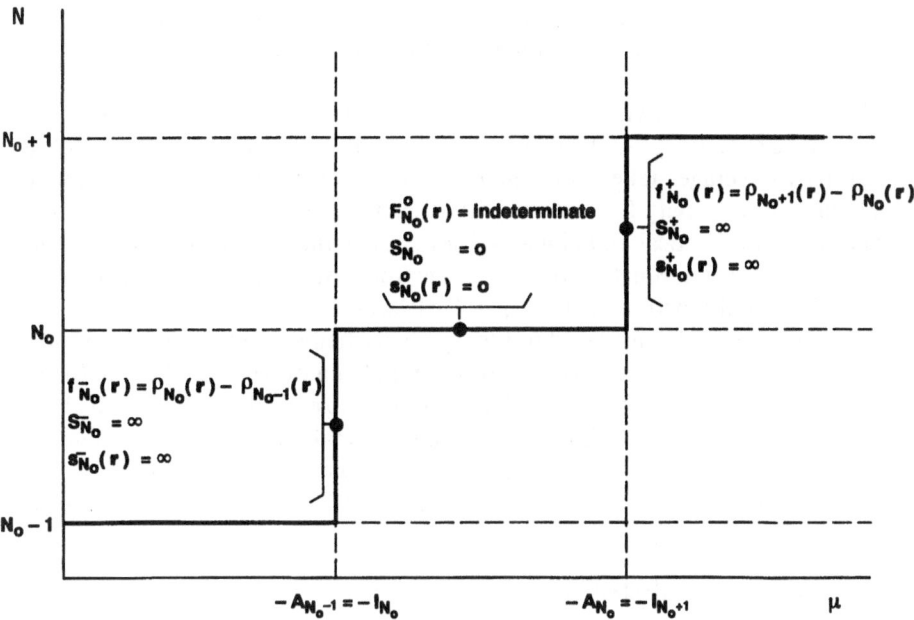

Fig. 1. The $\mathcal{N}$–μ staircase for $\mathcal{N}$ in the vicinity of the integer N_0. The values of the reactivity indices holding on the $\mathcal{N} = N_0$ tread and the two adjacent risers are indicated. These values are independent of the location of the $(\mu, \mathcal{N})$ point on the tread or on the risers. Here, $\mathcal{N}$ is the ensemble-average electron number and μ is the chemical potential

We now evaluate the reactivity indices S, f(r), s(r) as follows

$$S = \left[\frac{\partial \mathcal{N}}{\partial \mu} \right]_v = \sum_{N_0} \delta(\mu + A_{N_0}) = \sum_{N_0} \delta(\mu + I_{N_0}) \tag{33}$$

$$f(\mathbf{r}) = \left[\frac{\partial \rho(\mathbf{r})}{\partial \mathcal{N}} \right]_v = \left[\rho_{N_0+1}(\mathbf{r}) - \rho_{N_0}(\mathbf{r}) \right], \quad N_0 < \mathcal{N} < N_0 + 1 \tag{34a}$$

$$= \infty \qquad \mathcal{N} = N_0 \tag{34b}$$

$$s(\mathbf{r}) = \left[\frac{\partial \rho(\mathbf{r})}{\partial \mu} \right] = \sum_{N_0} \left[\rho_{N_0+1}(\mathbf{r}) - \rho_{N_0}(\mathbf{r}) \right] \delta(\mu + I_{N_0+1}) . \tag{35}$$

Since we know the complete, explicit dependence of all quantities on $\mathcal{N}$, it is not necessary to use the superscripts $\pm$. To make contact with the discussion of Sects. 2 and 3, however, we display in Fig. 1 on the riser below the $\mathcal{N} = N_0$ tread of the $\mathcal{N}$–μ staircase the nucleophilic quantities following from Eqs. (33)–(35) which correspond to the – reactive indices of Sects. 2 and 3, and on the riser above those corresponding to the + or electrophilic indices. On the tread in between, corresponding to fixed $\mathcal{N} = N_0$ and fixed $\rho(\mathbf{r}) = \rho_{N_0}(\mathbf{r})$, we have displayed the corresponding isoelectronic reactivity indices with superscript zero.

4.2 Inutility of the Softnesses and Discrepancy Between the Ensemble and the Kohn-Sham Fukui Functions for Localized Systems

The display of reactivities in Fig. 1 illustrates two points clearly. First, the definitions of Eqs. (33)–(35), taken from Parr and coworkers [8–12] but applied directly to averages taken with the ensemble of Eq. (28a), do not yield useful measures of reactivity for systems with discrete spectra, i.e., localized systems. The values of the local and global softnesses are either zero or infinity, and the electrophilic Fukui function of the N_0-electron system is identical to the nucleophilic Fukui function of the $(N_0 + 1)$-electron system.

Second, the results displayed in Fig. 1 only partially correspond to the results obtained from the Kohn-Sham equations in [2] and discussed in Sect. 3. In [2] we were careful to restrict the use of the softnesses to extended systems, which we discuss below. In Sect. 3, however, we did not make that restriction, and, in preparation for the present section, we presented our results for the softnesses as though they held for all systems. Now the local KS density of states can be expressed as

$$g(\mathbf{r}, \mu) = \sum_{is} n_i |\psi_i(\mathbf{x})|^2 \delta(\varepsilon_i - E) \tag{36}$$

in the ensemble version of the KS theory [18], in which the ε_i are the Kohn-Sham eigenvalues,

$$H_{KS} \psi_i = \varepsilon_i \psi_i , \tag{37}$$

and the n_i are the occupation numbers of the Kohn-Sham energy levels,

$$n_i = 1, \quad \varepsilon_i < \mu , \tag{38a}$$

$$0 < n_i < 1, \quad \varepsilon_i = \mu , \tag{38b}$$

$$n_i = 0, \quad \varepsilon_i > \mu , \tag{38c}$$

$$\sum_i n_i = \mathcal{N} . \tag{38d}$$

The case of nonintegral $\mathcal{N}$ and therefore of partial occupancy of the HOMO, Eq. (38b), occurs when $N_0 < \mathcal{N} < N_0 + 1$ and $\varepsilon_{N_0+1} = \mu = -I_{N_0+1}$ [18]. Thus the electrophilic and nucleophilic softnesses relate to situations in which $g(\mathbf{r}, \mu)$ diverges according to Eq. (36), and the isoelectronic softness relates to situations in which $g(\mathbf{r}, \mu)$ vanishes. The softnesses are correspondingly infinite or zero in the KS formulation of Sect. 3, which coincides precisely with the results of the previous subsection. There is no discrepancy between the two approaches in yielding softnesses which are useless for chemical consideration.

On the other hand, there is a discrepancy between the predictions of the Kohn-Sham theory of Sect. 2 and of the direct ensemble approach of the present section for the electrophilic and nucleophilic Fukui functions and for the chemical potential. Direct use of the ensemble of Eqs. (28) yields results for $f^\pm(\mathbf{r})$, Eq. (34a), and for μ, Eq. (30a), which are independent of $\mathcal{N}$ in the range

$N_0 < \mathcal{N} < N_0 + 1$. That this is not the case for the KS theory of Sect. 2 and [2] can be seen as follows. The electron density $\rho(\mathbf{r})$ must depend on $\mathcal{N}$, and indeed $f^{\pm}(\mathbf{r}) = 0$ is its derivative with respect to $\mathcal{N}$. Therefore, $V_{KS}(\mathbf{r})$, a nonconstant functional of $\rho(\mathbf{r})$, must also depend on $\mathcal{N}$ as, consequently, must H_{KS}. The eigenfunctions $\psi_i(x)$ and eigenvalues ε_i must in turn depend on $\mathcal{N}$. Indeed, the difference between $f^{\pm}(\mathbf{r})$, Eq. (20), and $\rho_{F}^{\pm}(\mathbf{r})$, Eq. (4), is the derivative of $\sum_{i=1s}^{N_0} |\psi_i(x)|^2$ with respect to $\mathcal{N}$, Eq. (3). Similarly, the Kohn-Sham Green's function is $\mathcal{N}$-dependent. Thus all the quantities entering $f^{\pm}(\mathbf{r})$ in Eq. (20) are $\mathcal{N}$-dependent. Furthermore, $\mu = \varepsilon_{N_0+1}$ holds in this range of $\mathcal{N}$, and μ is consequently $\mathcal{N}$-dependent.

4.3 Breakdown of the Ensemble Version of the Hohenberg-Kohn Theorem for Localized Systems

The results following from the use of the ensemble of Eq. (28) are rigorous; $D_{\mathcal{N}}$ is simply the $T = 0$ limit of the grand-canonical ensemble. The above discrepancies imply that the ensemble version of Kohn-Sham theory [18] becomes fundamentally incorrect at $T = 0$. The fault actually lies deeper. The ensemble version of the Hohenberg-Kohn theorem [18] on which Kohn-Sham theory is based does not hold at $T = 0$. The density functional on which the ensemble KS theory is based consequently does not exist at $T = 0$. The proof of the breakdown of the Hohenberg-Kohn theorem follows.

The grand potential Ω of the GCE is defined as

$$\Omega = -k_B T \ln Z \tag{39a}$$

$$Z = \mathrm{tr}\, e^{-\beta[H - \mu \hat{N}]}, \tag{39b}$$

where $\hat{N}$ is the electron number operator and k_B Boltzmann's constant. Omitting the internuclear Coulomb interaction in H for simplicity, we can write

$$H - \mu \hat{N} = H_e + \int d\mathbf{r}\, \hat{\rho}(\mathbf{r}) u(\mathbf{r}) . \tag{40}$$

In Eq. (40), H_e is the universal part of the electron Hamiltonian,

$$H_e = T_e + V_{ee} , \tag{41}$$

where T_e is the electronic kinetic energy and V_{ee} the Coulomb interaction between the electrons, and $\hat{\rho}(\mathbf{r})$ is the electron-density operator,

$$\int d\mathbf{r}\, \hat{\rho}(\mathbf{r}) = \hat{N} . \tag{42}$$

Thus the quantity

$$u(\mathbf{r}) = v(\mathbf{r}) - \mu \tag{43}$$

uniquely characterizes the system by specifying $H - \mu \hat{N}$. The grand potential Ω is then a functional of $u(\mathbf{r})$

$$\Omega = \Omega[u] . \tag{44}$$

The proof of the Hohenberg-Kohn theorem proceeds by establishing a one-to-one invertible map between $\rho(\mathbf{r})$ and u(r) so that Ω is shown to be a functional of $\rho(\mathbf{r})$ [18], the density functional

$$\Omega = \Omega[\rho] . \tag{45}$$

The Kohn-Sham equations are then derived using the variational principle satisfied by $\Omega[\rho]$ [18].

This sequence of arguments is correct at finite temperature, but there is no one-to-one invertible map between u(r) and $\rho(\mathbf{r})$ at absolute zero. When μ has the value $-I_{N_0+1}$, $\rho(\mathbf{r})$ can take on the continuously infinite set of values

$$\rho(\mathbf{r}) = (1 - \varepsilon)\, \rho^0_{N_0}(\mathbf{r}) + \varepsilon\rho^0_{N_0+1}(\mathbf{r}) , \tag{46a}$$

$$\mathcal{N} = N_0 + \varepsilon, \quad 0 < \varepsilon < 1 \tag{46b}$$

for fixed v(r). There is thus a one-to-many mapping between u(r) and $\rho(\mathbf{r})$ which does not meet the requirements for proof of the Hohenberg-Kohn theorem [18]. Moreover, for $N_0 < \mathcal{N} < N_0 + 1$, Ω takes on the constant value

$$\Omega = E^0_{N_0} + I_{N_0+1}N_0 , \tag{47}$$

independent of electron density.

Thus the density functional $\Omega[\rho]$, Eq. (45), does not exist at $T = 0$, and one cannot construct a Kohn-Sham potential v_{KS}. This can be understood in the following way. The physical system that the ensemble describes is a physical mixture of distinct, noninteracting N_0-and $(N_0 + 1)$-electron systems. The Kohn-Sham equations, were they to exist, would collapse these into a single pseudosystem with a nonintegral number of electrons. The notion of fractional occupancy, Eqs. (38), within the Kohn-Sham equations at $T = 0$ is simply incorrect at $T = 0$ for a nondegenerate HOMO [42].

Instead of supposing there to be a single Kohn-Sham potential, one can think of it as a vector in Fock space. For each sheet $\hat{N} = N$ of the latter, there is a component $v_{KS}(\mathbf{r}, N)$ and a corresponding set of Kohn-Sham equations. Density functional theory and Kohn-Sham theory hold separately on each sheet. Ensemble-average properties are then composed of weighted contributions from each sheet, computable sheet by sheet via the techniques of DFT and the KS equations. Nevertheless, though completely valid, this procedure would yield for the reactivity indices f(r), s(r), and S the results already obtained directly from Eqs. (28). We are left without proper definitions of chemical-reactivity indices for systems with discrete spectra at $T = 0$ [43].

4.4 Validity of the Theory for Extended Systems

Making the temperature finite restores DFT in its conventional form [18] but leads to reactivities which are violently temperature-dependent for $k_B T \ll$ relevant energy separations for systems with discrete spectra. All is not lost, however. There are systems, gapless systems, for which the Fukui function and the local and global softnesses have the meaning originally intended. These systems, which are extended in at least one dimension, have a dimension or fractal dimension of at least unity, and have a sufficiently dense distribution of states at and above the ground-state energy. For such systems, the thermodynamic average of a physical quantity taken in the GCE at low but nonvanishing temperatures is a temperature-independent, continuous, differentiable function of the ensemble-average number $\mathcal{N}$ provided the following five conditions are met [44]:

1. The mean spacing of energy levels $d_N(E)$ above the ground-state energy E_{N^0} should be exponentially small in N for all relevant N.

2. $k_B T$, while very small, should still be large enough that the energy scale can be divided into a set of adjacent energy intervals γ_N which satisfy

$$d_N(E) \ll \gamma_N \ll k_B T \qquad (48)$$

for all relevant N.

3. Averages of energy-level-dependent physical quantities over the energy intervals γ_N should then vary slowly with the central energy of each interval.

4. The quantities

$$\Omega_N = E_N^0 - \mu N \qquad (49)$$

have finite differences of order n, $\Delta^{(n)} \Omega_N$, which satisfy the condition

$$\beta \langle \Delta^{(n)} \Omega_N (N - \mathcal{N})^n \rangle \ll 1, \quad n > 2 , \qquad (50)$$

where the brackets indicate an average over the GCE. The requirement (50) is satisfied up to temperatures algebraically large in $\mathcal{N}$ if

$$(k_B T)^{\left(\frac{n}{2}-1\right)} \Delta^{(n)} \Omega_N / (\Delta^{(2)} \Omega_N)^{\frac{n}{2}} \ll 1, \quad n > 2 . \qquad (51)$$

The condition (51), a weak extensive property, is well satisfied for the relevant N by all extended systems of which we are aware. Although it is a sufficient condition, it may not be necessary.

5. The chemical potential μ is given by

$$\Delta^{(1)} \Omega_{N^*} = 0 , \qquad (52a)$$

$$\Delta^{(1)} E_{N^*}^0 = \mu , \qquad (52b)$$

$$|\mathcal{N} - N^*| \leq \tfrac{1}{2} , \qquad (52c)$$

where N* is the closest integer to $\mathcal{N}$.

We have shown [44] that when these conditions are met, not only do all discontinuities disappear, but all thermodynamic averages are smooth, temperature-independent functions of μ and $\mathcal{N}$ at sufficiently low temperatures, low but not so low as to violate Eq. (48). μ is a smooth function of $\mathcal{N}$, and all of the difficulties associated with those discrete spectra for which Eqs. (48) through (52) are violated simply disappear, as does the distinction between the electrophilic (+) and the nucleophilic (−) reactivities. Moreover, the notion of a Fock-space vector $v_{KS}(\mathbf{r}, \mathcal{N}, [\rho])$ for the Kohn-Sham potential introduced above can be used to derive the usual ensemble version of the Kohn-Sham theory [18], now seen to be valid only for systems which are gapless in the above sense.

We thus have well-defined charge-transfer reactivities (electrophilic/electrophobic) for gapless systems: the Fukui function,

$$f(\mathbf{r}) = \frac{\int d\mathbf{r}' K^{-1}(\mathbf{r}, \mathbf{r}') \sum_s \overline{|\psi_i(\mathbf{x}')|^{2^\mu}}}{\int d\mathbf{r} d\mathbf{r}' K^{-1}(\mathbf{r}, \mathbf{r}') \sum_s \overline{|\psi_i(\mathbf{x}')|^{2^\mu}}} , \tag{53a}$$

$$\sum_s \overline{|\psi_i(\mathbf{x}')|^{2^\mu}} = \frac{g(\mathbf{r}', \mu)}{D(\mu)} , \tag{53b}$$

$$D(\mu) = \int d\mathbf{r}\, g(\mathbf{r}, \mu) , \tag{53c}$$

$$f(\mathbf{r}) = s(\mathbf{r})/S ; \tag{53d}$$

the local softness

$$s(\mathbf{r}) = \int d\mathbf{r}' K^{-1}(\mathbf{r}, \mathbf{r}') g(\mathbf{r}', \mu) ; \tag{54}$$

and the global softness

$$S = \int d\mathbf{r}\, s(\mathbf{r}) . \tag{55}$$

It can be seen from Eq. (53a) that the average over states at the chemical potential of the KS orbital density plays the role of the Fukui frontier-orbital density,

$$\rho_F(\mathbf{r}) = \overline{\sum_s |\psi_i(\mathbf{x})|^{2^\mu}} . \tag{56a}$$

However, because of the screening action of the kernel K^{-1} [2], it is a rescaled frontier-orbital density

$$\tilde{\rho}_F(\mathbf{r}) = \frac{\rho_F(\mathbf{r})}{\int d\mathbf{r} d\mathbf{r}' K^{-1}(\mathbf{r}, \mathbf{r}') \rho_F(\mathbf{r})} \tag{56b}$$

which enters $f(\mathbf{r})$,

$$f(\mathbf{r}) = \int d\mathbf{r}' K^{-1}(\mathbf{r}, \mathbf{r}') \tilde{\rho}_F(\mathbf{r}') . \tag{56c}$$

The norm of $\rho_F(\mathbf{r})$ is unity,

$$\int d\mathbf{r} \rho_F(\mathbf{r}) = 1 , \tag{57}$$

but we expect that of $\tilde{\rho}_F(\mathbf{r})$ to be less than unity, as discussed in [2]. Isoelectronic reactivities have, however, disappeared into the quasicontinuum.

In conclusion, the analysis of the present section raises two further issues: How are electron-transfer reactivities to be defined for systems with discrete spectra or localized systems? How are the isoelectronic reactivities to be defined for both localized and extended systems? Before we address these more difficult questions, an issue which can be dealt with in the present limited framework for extended, i.e., gapless, sysems is discussed in the following section.

5 Local Nuclear Reactivities, Gapless Systems

Nakatsuji [45–47] obtained interesting interrelations of changes in electron density and nuclear configuration in a variety of contexts. He has shown, for example, that the centroid of the change in electron density tends to lag the change in nuclear coordinates in movement away from a stable configuration and to lead it in movement away from an unstable one. Among the configuration changes considered by Nakatsuji and his predecessors (cf. refs. cited by Nakatsuji [45–47]) were those associated with chemical reactions. This leads us to our fourth issue. Changes in nuclear configuration define chemical reactions. It is therefore necessary to define nuclear-reactivity indices and to expose clearly the interrelation between the nuclear and the electronic reactivities, as Nakatsuji did for the centroid of the electron density.

This issue was addressed in [2] for electron-transfer reactivities. The nuclear Fukui functions and softnesses defined there are valid for gapless systems. The analysis begins with nuclear forces specified through the Feynman-Hellmann theorem [48],

$$\mathbf{F_a} = \sum_b' \frac{\partial}{\partial \mathbf{R_a}} \left(\frac{Z_a Z_b e^2}{|\mathbf{R_a} - \mathbf{R_b}|} \right) + \int d\mathbf{r} \frac{\partial}{\partial \mathbf{R_a}} \left(\frac{Z_a e^2}{|\mathbf{R_a} - \mathbf{r}|} \right). \tag{58}$$

In Eq. (58), the indices a and b specify the nuclei, $\mathbf{F_a}$ is the force on nucleus a, Z_a its charge, and $\mathbf{R_a}$ its position. We introduced in [2], in analogy to the electron-transfer reactivities, the nuclear Fukui function

$$\phi_a = \left[\frac{\partial \mathbf{F_a}}{\partial \mathcal{N}} \right]_v \tag{59}$$

and the nuclear softness,

$$\sigma_a = \left[\frac{\partial \mathbf{F_a}}{\partial \mu} \right]_v. \tag{60}$$

From these definitions, it follows immediately that the nuclear Fukui function is

$$\phi_a = \int d\mathbf{r} \frac{\partial}{\partial \mathbf{R_a}} \left(\frac{Z_a e^2}{|\mathbf{R_a} - \mathbf{r}|} \right) f(\mathbf{r}) \tag{61}$$

and the nuclear softness is

$$\sigma_a = \int d\mathbf{r} \frac{\partial}{\partial \mathbf{R_a}} \left(\frac{Z_a e^2}{|\mathbf{R_a} - \mathbf{r}|} \right) s(\mathbf{r}). \tag{62}$$

A remarkable pair of equations then results from substitution of Eqs. (56c) and (54) into Eqs. (61) and (62), respectively,

$$\phi_a = \int d\mathbf{r} F_a(\mathbf{r}) \tilde{\rho}_F(\mathbf{r}), \tag{63}$$

$$\sigma_a = \int d\mathbf{r} \, F_a(\mathbf{r}) \, g(\mathbf{r}, \mu) \,, \tag{64}$$

where $F_a(\mathbf{r})$ is the force resulting from the screening of the bare nuclear force by the KSPRF $\mathcal{K}(\mathbf{r}, \mathbf{r}')$ [2],

$$F_a(\mathbf{r}) = \int d\mathbf{r} \mathcal{K}^{-1}(\mathbf{r}, \mathbf{r}') \frac{\partial}{\partial \mathbf{R}_a} \left(\frac{Z_a e^2}{(\mathbf{r}' - \mathbf{R}_a)} \right). \tag{65}$$

Equations (63) and (64) resolve issue four. Nuclear reactivity indices have been defined and clearly display the drivers of the nuclear response to transfers of electrons into and out of the system. The rescaled frontier-orbital density $\tilde{\rho}_F(\mathbf{r})$ drives that response through the screened force, Eq. (65), when $\mathcal{N}$ is changed directly, Eq. (63), and the local DOS $g(\mathbf{r}, \mu)$ drives it when μ is changed, Eq. (64).

6 Chemical Reactivity as a Nonlocal Response to a Nonlocal Perturbation

Thus far we have considered only the quantities $f(\mathbf{r})$ and $s(\mathbf{r})$ as indices of chemical reactivity. These quantities, however, are local responses to the global perturbations $\partial \mathcal{N}$ and $\partial \mu$, respectively. Chemical reactions proceed by nonlocal responses of one reactant to nonlocal perturbations generated in chemical attack by another reactant. Thus, a fifth issue emerges, how to define nonlocal reactivity indices. Berkowitz and Parr (BP) [15] addressed this issue by introducing the softness kernel,

$$s^\mu(\mathbf{r}, \mathbf{r}') = -\frac{\delta\rho(\mathbf{r})}{\delta u(\mathbf{r}')}. \tag{66}$$

We have added the superscript μ because one can show by direct consideration of the GCE that

$$s^\mu(\mathbf{r}, \mathbf{r}') = -\left[\frac{\delta\rho(\mathbf{r})}{\delta v(\mathbf{r}')} \right]_\mu. \tag{67}$$

Maintaining constancy of μ implies that $s^\mu(\mathbf{r}, \mathbf{r}')$ is an electron-transfer reactivity index. In contrast to BP, who assumed implicitly that the definition of Eq. (66) was meaningful for all systems at all temperatures, we restricted in [3] consideration of the softness kernel to gapless systems for the reasons presented in Sect. 4. As a consequence, we were able in [3] to obtain from Eq. (66) an explicit form for $s^\mu(\mathbf{r}, \mathbf{r}')$ via the Kohn-Sham equations,

$$s^\mu(\mathbf{r}, \mathbf{r}') = \int d\mathbf{r}'' K^{-1}(\mathbf{r}, \mathbf{r}'') \, P(\mathbf{r}'', \mathbf{r}') \,. \tag{68a}$$

The charge-transfer softness kernel has many very interesting properties, of which we cite several in the following. First, it is real and symmetric,

$$s^\mu(\mathbf{r}, \mathbf{r}'') = s^\mu(\mathbf{r}', \mathbf{r}) . \tag{68b}$$

BP give a general proof that its integral is the local softness,

$$\int d\mathbf{r}' s^\mu(\mathbf{r}, \mathbf{r}') = s(\mathbf{r}) . \tag{69}$$

Equation (69) also follows readily from Eq. (68a) by consideration of Eqs. (54), (23), and (25). Nalewajski [19, 21] introduced what we term the isoelectronic softness kernel. We define it here as a functional derivative of the electron density,

$$s^{\mathcal{N}}(\mathbf{r}, \mathbf{r}') = -\left[\frac{\delta\rho(\mathbf{r})}{\delta v(\mathbf{r}')}\right]_{\mathcal{N}}, \tag{70}$$

which can be defined for systems with discrete as well as quasicontinuous spectra. For N equal to an integer N_0, it has the explicit form

$$s^{\mathcal{N}}(\mathbf{r}, \mathbf{r}') = \sum_n{}' \frac{(\hat{\rho}(\mathbf{r}))_{on}(\hat{\rho}(\mathbf{r}'))_{no} + (\hat{\rho}(\mathbf{r}'))_{on}(\hat{\rho}(\mathbf{r}))_{no}}{E_{no}}, \tag{71a}$$

$$H_{N_0}\psi_n = E_n\psi_n , \tag{71b}$$

$$(\hat{\rho}(\mathbf{r}))_{on} = (\psi_0, \hat{\rho}(\mathbf{r})\psi_n) , \tag{71c}$$

In Eq. (71b), H_{N_0} is the N_0-particle Hamiltonian, the subscript 0 specifies ground-state properties, $\hat{\rho}(\mathbf{r})$ is the electron-density operator, and $E_{no} = E_n - E_0$.

BP [15] prove by thermodynamic arguments which imply the use of the GCE that

$$s^\mu(\mathbf{r}, \mathbf{r}') = s^{\mathcal{N}}(\mathbf{r}, \mathbf{r}') + s(\mathbf{r})s(\mathbf{r}')/S . \tag{72a}$$

We shall refer to Eq. (72) as the BP relation. BP presumed their proof to be valid for all systems. It does not hold for localized systems at $T = 0$ when $\mu = I_{N_0+1}$ and $N_0 < \mathcal{N} < N_0 + 1$ because $\rho(\mathbf{r})$ is then indeterminate at fixed μ and the definition of Eq. (66) loses its meaning. However, an alternative proof has been constructed for gapless systems using the GCE along the lines of the arguments outlined in Sect. 4 [42, 44]. If one uses the definition of Eq. (67) instead of Eq. (66), one can show that the BP relation holds for *all* systems at $T = 0$, provided that

$$\varepsilon_{N_0} < \mu < \varepsilon_{N_0} + 1 \tag{72b}$$

is maintained for the localized systems so that $\mathcal{N}$ remains constant at the integer N_0. Recognizing that $s(\mathbf{r})$ then vanishes for localized systems, one sees that Eq. (72a) reduces to

$$s^\mu(\mathbf{r}, \mathbf{r}') = s^{\mathcal{N}}(\mathbf{r}, \mathbf{r}') . \tag{72c}$$

What one has done here is to recapture the BP relation for localized systems at the cost of avoiding a proper definition of charge-transfer chemical reactivities for such systems. Equation (72c) implies that we have introduced only isoelectronic reactivities.

The analysis of [3] which leads to Eq. (68a) is also valid for localized systems for which it yields the result

$$s^{\mu}(\mathbf{r}, \mathbf{r}') = s^{\mathscr{N}}(\mathbf{r}, \mathbf{r}') = \int d\mathbf{r}'' \mathscr{X}^{-1}(\mathbf{r}, \mathbf{r}'') \, P(\mathbf{r}'', \mathbf{r}') \, . \tag{73}$$

From Eqs. (73a), (23), (25), and (72b), it follows that

$$s(\mathbf{r}) = \int d\mathbf{r}' s^{\mu}(\mathbf{r}, \mathbf{r}') = 0 \, , \tag{74}$$

as it should be for localized systems. In fact, the restriction of Eq. (72b) alone, guaranteeing that $\mathscr{N}$ is constant at N_0, is sufficient to impose Eqs. (72c) and (74).

Obtaining a Kohn-Sham expression for $s^{\mathscr{N}}(\mathbf{r}, \mathbf{r}')$ for gapless systems is most conveniently effected through the BP relation of Eq. (72a); the result is

$$s^{\mathscr{N}}(\mathbf{r}, \mathbf{r}') = \int d\mathbf{r}'' L^{-1}(\mathbf{r}, \mathbf{r}'') Q(\mathbf{r}'', \mathbf{r}') \, , \tag{75a}$$

$$L(\mathbf{r}, \mathbf{r}') = \delta(\mathbf{r}, \mathbf{r}') + \int d\mathbf{r}'' Q(\mathbf{r}, \mathbf{r}'') \, W(\mathbf{r}'', \mathbf{r}') \, , \tag{75b}$$

$$Q(\mathbf{r}, \mathbf{r}') = P(\mathbf{r}, \mathbf{r}') - \frac{g(\mathbf{r}, \mu) \, g(\mathbf{r}', \mu)}{D(\mu)} \, . \tag{75c}$$

By comparing Eq. (75) with Eqs. (68a) and (23), one sees that $s^{\mathscr{N}}(\mathbf{r}, \mathbf{r}')$ has the same formal structure as $s^{\mu}(\mathbf{r}, \mathbf{r}')$ but with $P(\mathbf{r}, \mathbf{r}')$ replaced everywhere by $Q(\mathbf{r}, \mathbf{r}')$. Equation (75) shows that $Q(\mathbf{r}, \mathbf{r}')$ is the polarization propagator for constant $\mathscr{N}$, i.e., $P(\mathbf{r}, \mathbf{r}')$ is defined through (24a), whereas

$$Q(\mathbf{r}, \mathbf{r}') = \left[\frac{\delta \rho(\mathbf{r})}{\delta v_{KS}(\mathbf{r}')} \right]_{\mathscr{N}} \, . \tag{76}$$

Finally, instead of Eq. (25),

$$\int d\mathbf{r}' Q(\mathbf{r}, \mathbf{r}') = 0 \tag{77}$$

holds.

We note that in the course of dealing with the fifth issue, the introduction of nonlocal chemical responses, we have successfully addressed the third issue, the definition of isoelectronic reactivity indices for both localized and extended systems. We note also that recognition of the need for a nonlocal description of chemical responses came early. Hückel's bond-bond polarizability [49] is in fact a simplified version of $P(\mathbf{r}, \mathbf{r}')$ [3].

We note that there is an important and useful relation between the static dielectric functions $\varepsilon_\alpha(\mathbf{r}, \mathbf{r}')$ and the softness kernels $s^\alpha(\mathbf{r}, \mathbf{r}')$ [3],

$$\varepsilon_\alpha^{-1}(\mathbf{r}, \mathbf{r}') = \delta(\mathbf{r}, \mathbf{r}') - \int d\mathbf{r}'' \frac{e^2}{|\mathbf{r} - \mathbf{r}''|} s^\alpha(\mathbf{r}, \mathbf{r}') \, . \tag{78}$$

In Eq. (78), we have added the subscript $\alpha = \mu, \mathscr{N}$ to distinguish between dielectric responses at constant μ and constant $\mathscr{N}$, a distinction not usually made in dielectric-response theory. That it should be made is obvious from the definition of the dielectric function,

$$\varepsilon_\alpha^{-1}(\mathbf{r}, \mathbf{r}') = \left[\frac{\delta v_H(\mathbf{r})}{\delta v(\mathbf{r}')} \right]_\alpha , \tag{79a}$$

$$v_H(\mathbf{r}) = v(\mathbf{r}) + \int d\mathbf{r}' \frac{e^2}{|\mathbf{r} - \mathbf{r}'|} \rho(\mathbf{r}') . \tag{79b}$$

Finally, just as we were able to define in Sect. 5 the nuclear softness, a local nuclear reactivity index in analogy with the electronic local softness, we can now introduce nonlocal nuclear reactivity indices in analogy to the electronic softness kernels [3]. We define the nuclear softness kernels as follows

$$\sigma_a^\alpha(\mathbf{r}) = -\left[\frac{\delta F_a}{\delta v(\mathbf{r})} \right]_\alpha . \tag{80}$$

It follows from the Feynman-Hellmann theorem, Eq. (58), that

$$\sigma_a^\alpha(\mathbf{r}) = \frac{\partial}{\partial \mathbf{R}_a} \int d\mathbf{r}' \frac{Z_a e^2}{|\mathbf{R}_a - \mathbf{r}'|} s^\alpha(\mathbf{r}', \mathbf{r}) . \tag{81}$$

This in turn implies from Eqs. (68a) and (75) that

$$\sigma_a^\alpha(\mathbf{r}) = \int d\mathbf{r}' F_a^\alpha(\mathbf{r}') P^\alpha(\mathbf{r}', \mathbf{r}) . \tag{82}$$

In Eq. (82) we have introduced a more compact notation through consistent use of the superscript or subscript α:

$$F_a^\alpha(\mathbf{r}) = \int d\mathbf{r}' \mathcal{K}_\alpha^{-1}(\mathbf{r}, \mathbf{r}') \frac{\partial}{\partial \mathbf{R}_a} \frac{Z_a e^2}{|\mathbf{r}' - \mathbf{R}_a|} , \tag{83a}$$

$$\mathcal{K}_\alpha(\mathbf{r}, \mathbf{r}') = K_\alpha(\mathbf{r}, \mathbf{r}')^T , \tag{83b}$$

$$K_\alpha(\mathbf{r}, \mathbf{r}') = \delta(\mathbf{r}, \mathbf{r}') + \int d\mathbf{r}'' P^\alpha(\mathbf{r}, \mathbf{r}'') W(\mathbf{r}'', \mathbf{r}') , \tag{83c}$$

$$K_\mu(\mathbf{r}, \mathbf{r}') \equiv K(\mathbf{r}, \mathbf{r}') , \tag{83d}$$

$$K_{\mathcal{N}}(\mathbf{r}, \mathbf{r}') \equiv L(\mathbf{r}, \mathbf{r}') , \tag{83e}$$

$$P^\mu(\mathbf{r}, \mathbf{r}') \equiv P(\mathbf{r}, \mathbf{r}') , \tag{83f}$$

$$P^{\mathcal{N}}(\mathbf{r}, \mathbf{r}') \equiv Q(\mathbf{r}, \mathbf{r}') . \tag{83g}$$

Returning to Eq. (82), we see that $P^\alpha(\mathbf{r}, \mathbf{r}')$ is the driving force for the nonlocal nuclear response, Eq. (80).

The relation between the softness kernels and the dielectric functions, Eq. (78), allows one to transform Eq. (81) to

$$\sigma_a^\alpha = Z_a \frac{\partial}{\partial \mathbf{R}_a} [\delta(\mathbf{R}_a - \mathbf{r}) - \varepsilon_\alpha^{-1}(\mathbf{R}_a, \mathbf{r})] , \tag{84}$$

an expression of use in relation to the dynamics of nuclear motion [3].

7 The Ambiguity of the Frontier-Orbital Concept and the Complexity of Chemical Responses

We have now seen that the effort of Parr and collaborators [8–12] to put Fukui's frontier-orbital concept of chemical reactivity on sound footing in density-functional theory through the definition of the Fukui function and the local and global softness works only for extended systems. This restriction to extended systems raises a sixth issue. In both the local softness and the Fukui function, Eqs. (54) and (53a), the orbitals at the chemical potential represent both the LUMO and the HOMO in the Fukui sense. However, there is a continuum of unoccupied KS states above the chemical potential accessible even to weak chemical perturbations any linear combination of which could in principle be selected as the LUMO, and similarly for states below μ and the HOMO. This ambiguity in the frontier-orbital concept obviously applies as well to localized systems when there is more than one KS state significantly affected by a chemical perturbation.

Fortunately, the reactivity kernels $s^\alpha(\mathbf{r}, \mathbf{r}')$ resolve this issue, as is clearly implied by the studies of $s^{\mathcal{N}}(\mathbf{r}, \mathbf{r}')$ of Nalewajski and coworkers [22–24] with their introduction of the normal modes of reactivity.

The reactivity kernels have a spectral representation

$$s^\alpha(\mathbf{r}, \mathbf{r}') = \sum_\gamma \chi^\alpha_{\gamma\alpha}(\mathbf{r}') \, s^\alpha_\gamma \chi_{\gamma\alpha}(\mathbf{r}) , \tag{85a}$$

$$(\chi_{\gamma\alpha}, \chi_{\gamma'\alpha}) = \delta_{\gamma\gamma'} . \tag{85b}$$

The infinite set of eigenfunctions of $s^\alpha(\mathbf{r}, \mathbf{r}')$, the $\chi_{\gamma\alpha}$, represents the softness modes. There is thus an infinite set of potential responses of the system to chemical attack. The corresponding infinite set of eigenvalues of $s^\alpha(\mathbf{r}, \mathbf{r}')$, the s^α_γ, constitutes the softness spectrum and measures the intensity of response of a given mode, while the distance dependence of the $\chi_{\gamma\alpha}(\mathbf{r})$ displays the geometry of the responses.

We have proved [3] that the softness spectrum is bounded,

$$0 \le s^\alpha_\gamma \le s^\alpha_M , \tag{86}$$

so that $\chi_{M\alpha}(\mathbf{r})$ is the softest, most reactive mode. Clearly, however, the chemical response must be determined by the chemical stimulus, and may well differ from $\chi_{M\alpha}(\mathbf{r})$. It almost certainly differs from $s(\mathbf{r})$, which is the response to a uniform stimulus, irrelevant except in certain electrochemical contexts. Chemical stimuli do not appear in the theory as we have presented it thus far, raising a seventh issue–how to define and introduce chemical stimuli.

Returning to the matter of the bounds of the softness spectrum [3], we have found it easier to address it through study of the hardness kernels [15], reciprocals to the softness kernels,

$$\eta^\alpha(\mathbf{r}, \mathbf{r}') = \tfrac{1}{2}(s^\alpha)^{-1}(\mathbf{r}, \mathbf{r}') , \tag{87}$$

which have the spectral representation

$$\eta^\alpha(\mathbf{r}, \mathbf{r}') = \sum_\gamma \chi_{\gamma\alpha}^*(\mathbf{r}') \, (1/2s_\gamma^\alpha) \, \chi_{\gamma\alpha}(\mathbf{r}) \;. \tag{88}$$

It is straightforward to show that [15]

$$\eta^\alpha(\mathbf{r}, \mathbf{r}') = \frac{1}{2} \left[\frac{\delta^2 E[\rho]}{\delta\rho(\mathbf{r})\delta\rho(\mathbf{r}')} \right]_\alpha . \tag{89}$$

The condition for stability of the ground state is that $E[\rho]$ be a convex functional and therefore that $\eta^\alpha(\mathbf{r}, \mathbf{r}')$, twice its Hessian, be positive-definite. $\eta^\alpha(\mathbf{r}, \mathbf{r}')$ thus possesses a minimum eigenvalue and $s^\alpha(\mathbf{r}, \mathbf{r}')$ a maximum, s_M^α. A perturbation theoretic analysis sketched briefly in [3] then establishes that the upper bound of the spectrum of $\eta^\alpha(\mathbf{r}, \mathbf{r}')$ is ∞, and therefore the lower bound of the s_γ^α is zero.

8 The Total Energy, Chemical Stimuli, and the Distinction Between Static and Dynamic Reactivities for Disjoint Systems

In Sect. 7, we raised the question of what were the chemical stimuli to which the reactivity indices defined in Sect. 6, the softness kernels, were presumed to be the responses, our seventh issue. Now there are various broad categories of reactions to be considered, unimolecular, bimolecular, and multimolecular. The former occur via thermal activation over a barrier, tunneling through the barrier, or some combination of both. There is no stimulus, and the softness kernels defined as responses of the electron density to changes in external or nuclear potential are irrelevant. For the study of unimolecular reactions, one needs only information about the total energy in the relevant configuration space of the molecule.

The softness kernels are relevant to the remaining cases of two or more interacting systems. However, they do not by themselves provide sufficient information to constitute a basis for a theory of chemical reactivity. Clearly, the chemical stimulus to one molecule in a bimolecular reaction is provided by the other. That being the case, an eighth issue arises. Both the perturbing system and the responding system have internal dynamics, yet the softness kernel is a static response function. Dynamic reactivities need to be defined.

Moreover, calculation of reaction rates, the ultimate goal of chemical-kinetic or -dynamic theory, requires knowledge of the total energy surfaces of the interacting complex. This raises a ninth issue, the final one to be considered in the present chapter: How do the reactivity indices relate to the energy of interaction of the reactants? Implicit in this question is the requirement that the interaction energy be expressible in terms of the properties of the isolated reactants, restricting analysis to second-order perturbation theory. Thus, strictly speaking, chemical-reactivity theory can only define the initial part of the reaction pathway, which we term the gateway or portal. Despite this limitation, the perturbation calculation of the interaction energy would settle successfully the seventh, eighth, and ninth issues. At present, however, the perturbation theory has been developed for nonoverlapping, i.e., disjoint systems [50], and we present the results in the following paragraphs together with discussion of the resolution of the seventh and eighth issues. The restriction to systems whose wave functions do not overlap eliminates the possibility of electron transfer and restricts us to isoelectronic reactivities, leaving the theory of chemical reactivity incomplete.

Nalewajski [19–21] formulated a theory of static reactivity kernels prior to but closely related to that of Berkowitz and Parr [15] which started out from a second-order perturbative treatment of the total energy. He identified an external static potential of unspecified origin with the chemical stimulus and made the first connection between softness kernels and the total energy. His result is, in our notation,

$$E^{(2)} = -\tfrac{1}{2}\iint dr\,dr'\,s''(r, r')V(r)V(r') . \tag{90}$$

Here, $V(r)$ is the unspecified external potential. The static isoelectronic softness kernel $s''(r, r')$ entering Nalewajski's result is the same as that defined in Eq. (71). The chemical stimulus is thus the nonlocal quantity $V(r) V(r')$ in Nalewajski's theory, but still unspecified.

We now carry out our program of defining chemical stimuli, introducing dynamic reactivities, and relating energies of interaction to the reactivity indices [50]. Considered the two weakly interacting, disjoint systems α and β shown in Fig. 2. By disjoint we mean that such wave-function products as $\psi_{m\alpha}(\cdots r\cdots)$

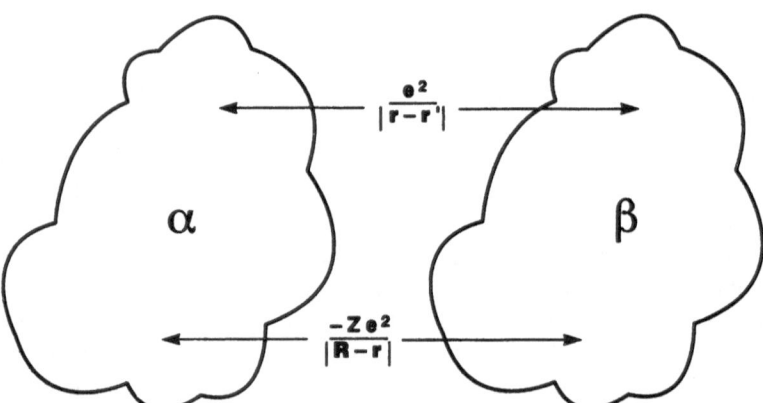

Fig. 2. The two weakly coupled, disjoint systems α and β perturb each other through the Coulomb interactions of the electrons of one with the electrons ($e^2/|r - r'|$) and the nuclei ($-Ze^2/|R - r|$) of the other

$\psi_{n\beta}(\cdots \mathbf{r} \cdots)$ are negligibly small for all $\mathbf{r}$. By weakly interacting we mean that perturbation theory need only be carried to second order in the Coulomb interactions of an electron in α with electrons and nuclei in β and vice versa. We obtain a result for the interaction energy E_{INT} which is standard in the theory of dispersion forces [51, 52] in one form or another, except that we have transcribed it into the language of reactivity theory,

$$E_{INT} = E_{INT}^{(1)} + E_{INT}^{(2)}. \tag{91}$$

The first-order term is simply the static Coulomb interaction between the unperturbed charge densities of the two systems, $V_{\alpha\beta}$,

$$E_{INT}^{(1)} = \int\int d\mathbf{r}\, d\mathbf{r}' \frac{n_\alpha(\mathbf{r})n_\beta(\mathbf{r}')e^2}{|\mathbf{r} - \mathbf{r}'|} = V_{\alpha\beta} . \tag{92}$$

In Eq. (92),

$$n_{\alpha,\beta} = \rho_{\alpha,\beta}(\mathbf{r}) - \sum_{a,b=1}^{N_{\alpha,\beta}} Z_{a,b}\delta(\mathbf{R}_{a,b} - \mathbf{r}) \tag{93}$$

is the charge density of α or β divided by (-e). Note that Eq. (92) implies that forces on the nuclei exist which are important for the early stages of the reaction even before the electronic density begins to change. In this early stage of the reaction, the nuclei of α always lead the electrons of β and similarly for the nuclei of β and the electrons of α, contrary to Nakatsuji's findings [45–47].

The second-order term is much richer and more complex. It contains a static and a dynamic part,

$$E_{INT}^{(2)} = E_{INT}^{(2S)} + E_{INT}^{(2D)} . \tag{94}$$

The static part is simply the energy of polarization of α by the static field of β and vice versa,

$$E_{INT}^{(2S)} = -\tfrac{1}{2}\int\int d\mathbf{r}\, d\mathbf{r}'\, s_\alpha''(\mathbf{r}, \mathbf{r}')V_\beta(\mathbf{r})V_\beta(\mathbf{r}') - \tfrac{1}{2}\int\int d\mathbf{r}\, d\mathbf{r}'\, s_\beta''(\mathbf{r}, \mathbf{r}')V_\alpha(\mathbf{r})V_\alpha(\mathbf{r}') . \tag{95}$$

In Eq. (95), $V_{\alpha,\beta}(\mathbf{r})$ is the static Coulomb potential energy produced at β, α by α, β,

$$V_{\alpha,\beta}(\mathbf{r}) = \int d\mathbf{r}' \frac{n_{\alpha,\beta}(\mathbf{r}')e^2}{|\mathbf{r} - \mathbf{r}'|} . \tag{96}$$

The dynamic part is simply the energy associated with the dispersion forces between α and β,

$$E_{INT}^{(2D)} = -\frac{1}{2\pi}\int_0^\infty du \int\int\int\int d\mathbf{r}_1 d\mathbf{r}_1' d\mathbf{r}_2 d\mathbf{r}_2'\, s_\alpha''(\mathbf{r}_1, \mathbf{r}_2; iu)$$

$$\times \frac{e^2}{|\mathbf{r}_1 - \mathbf{r}_1'|} \frac{e^2}{|\mathbf{r}_2 - \mathbf{r}_2'|} s_\beta''(\mathbf{r}_1', \mathbf{r}_2'; iu) . \tag{97}$$

167

The quantities $s_{\alpha,\beta}^{\mathscr{N}}(\mathbf{r}, \mathbf{r}'; iu)$ are the sought after dynamic reactivities needed to settle issue eight,

$$s^{\mathscr{N}}(\mathbf{r}, \mathbf{r}'; iu) = \sum_n{}' \frac{[(\hat{\rho}(\mathbf{r}))_{on}(\hat{\rho}(\mathbf{r}'))_{no} + (\hat{\rho}(\mathbf{r}'))_{on}(\hat{\rho}(\mathbf{r}))_{no}]}{(E_{no})^2 + u^2} E_{no}. \tag{98}$$

Comparing Eq. (98) for $s^{\mathscr{N}}(\mathbf{r}, \mathbf{r}'; iu)$ with the expression in Eq. (71a) for $s^{\mathscr{N}}(\mathbf{r}, \mathbf{r}')$, one sees that

$$s^{\mathscr{N}}(\mathbf{r}, \mathbf{r}') = s^{\mathscr{N}}(\mathbf{r}, \mathbf{r}'; iu)|_{u=0}. \tag{99}$$

An expression analogous to Eq. (70) can also be derived [50], but with the static variation $\rho_v(\mathbf{r}')$ replaced by the imaginary frequency component of a time-dependent variation,

$$s^{\mathscr{N}}(\mathbf{r}, \mathbf{r}'; iu) = -\left[\frac{\delta\rho(\mathbf{r}, iu)}{\delta v(\mathbf{r}', iu)}\right]_{\mathscr{N}}. \tag{100}$$

The dynamic softness is relatively unimportant when α and/or β are/ is charged. It can be of comparable importance to the static softness when α and/or β possess/es (a) dipole moment(s) and are/is polar. It can dominate the static softness when α and β are nonpolar. Its importance increases with increasing polarizability.

Clearly issue nine has been dealt with also. We have now expressed the energy of interaction of α and β in terms of their respective static and dynamic reactivities. What remains to discuss is how the concept of a chemical stimulus enters the theory through the structure of $E_{INT}^{(2)}$, settling issue seven. Each system presents both a static and dynamic stimulus to the other. The static stimulus which β presents to α is $V_\beta(\mathbf{r}) V_\beta(\mathbf{r}')$. It is intrinsically nonlocal. In this early stage of the reaction, the stimulus is to be visualized as occurring at pairs of points $\mathbf{r}$ and $\mathbf{r}'$ simultaneously. The dynamic interaction is not quite so neatly decomposed into a single stimulus-response product. It has the form of a semi-infinite integral over imaginary frequencies, i.e., over an infinite set of channels, of a stimulus-response product in each channel. Moreover, the systems simultaneously stimulate and respond to one another. The dynamic softness kernels act as both stimulus and response. The quantity

$$\iint d\mathbf{r}\, d\mathbf{r}'\frac{e^2}{|\mathbf{r}_1 - \mathbf{r}_1'|}\frac{e^2}{|\mathbf{r}_2 - \mathbf{r}_2'|} s_\beta^{\mathscr{N}}(\mathbf{r}_1', \mathbf{r}_2'; iu) \tag{101}$$

can be regarded as the stimulus of α by β and $s_\alpha^{\mathscr{N}}(\mathbf{r}_1, \mathbf{r}_2; iu)$ the response of α. Alternatively,

$$\iint d\mathbf{r}_1\, d\mathbf{r}_2 s_\alpha^{\mathscr{N}}(\mathbf{r}_1, \mathbf{r}_2)\frac{e^2}{|\mathbf{r}_1 - \mathbf{r}_1'|}\frac{e^2}{|\mathbf{r}_2 - \mathbf{r}_2'|} \tag{102}$$

can be regarded as the stimulus of β by α and $s_\beta^{\mathscr{N}}(\mathbf{r}_1', \mathbf{r}_2'; iu)$ the response of β. What is important is the relationship between total energy and reactivity indices.

9 Conclusion

In the preceding review of chemical-reactivity theory in Sects. 2 through 8, we raised nine issues which addressed specific weaknesses in the foundations of the theory and described the partial or complete resolution of each achieved to date. Those issues are: (1) What are the explicit and exact expressions for the electron-transfer reactivities defined by Parr and collaborators [8–12] (the Fukui function, the local softness, and the global softness) which follow from Kohn-Sham theory [7]? (2) These electron-transfer reactivities are meaningful only for extended systems with quasicontinuous or continuous spectra. How are electron-transfer reactivities to be defined for localized systems with discrete spectra? (3) How are isoelectronic reactivities to be defined for both localized and extended systems? (4) Changes in nuclear configuration define chemical reactions. Nuclear reactivities should therefore be introduced. What are the nuclear analogs of the Fukui function and the local softness for extended systems? (5) Chemical reactions entail nonlocal responses of one reactant to nonlocal perturbations generated in chemical attack by another. How should nonlocal indices of chemical reactivity be defined? (6) The frontier-orbital concept of Fukui is ambiguous. The HOMO and LUMO are ill-defined in extended systems, and more orbitals than simply the LUMO and the HOMO can participate in chemical reactions even in localized systems. How does one encompass in reactivity indices the full complexity of chemical responses? (7) The chemical reactivity indices have been defined as responses. What are the corresponding chemical stimuli? (8) The reactivity indices defined in the literature are all static response functions, yet the internal dynamics of the reactants must play a role in their chemical response. What are the corresponding dynamic reactivities? (9) It is the total energy of interaction of the reactants which is fundamental to our understanding of chemical reactions. How do the chemical-reactivity indices relate to the interaction energy?

We now summarize the present status of each of these issues:

1. Explicit KS expressions for the electron-transfer reactivities ($f^{\pm}(r)$, $s^{\pm}(r)$, $S^{\pm}$) defined by Parr and collaborators or implicit in their work [8–12] were obtained in [2] and reviewed in Sect. 3, cf. Eqs. (20) and (21). All quantities entering those expressions are in principle computable via Kohn-Sham theory, using any given approximation to the energy functional $F[\rho]$, Eq. (18b), and the exchange-correlation potential $v_{xc}(r)$, Eq. (19).

2. The electron-transfer reactivities are defined as derivatives of the electron-density $\rho(r)$ with respect to total electron number $\mathcal{N}$, $f^{\pm}(r)$, or chemical potential μ, $s^{\pm}(r)$. The treatment of $\mathcal{N}$ as a continuous variable [8–12] is justified by reference to the ensemble formulation of density-functional theory [8,18] and, in consequence, of the Kohn-Sham theory. We show in Sect. 4, in previously unpublished work [42], that this ensemble formulation yields either vanishing or infinite local and global softnesses for localized systems with

discrete electronic energy spectra at $T = 0$ and inconsistent results for the Fukui functions of those systems. Going to finite temperature eliminates these features but results in wildly temperature-dependent reactivity indices when $k_B T$ is much smaller than relevant energy gaps, a common situation. We show that the origin of these difficulties is the breakdown (at $T = 0$ for localized systems) of the ensemble version of the Hohenberg-Kohn theorem, on which density functional theory [16] and consequently Kohn-Sham theory [17] are based, because of the absence of an invertible map [18] between $u(\mathbf{r}) = v(\mathbf{r}) - \mu$ and $\rho(\mathbf{r})$. On the other hand, this breakdown does not occur for extended systems with quasicontinuous spectra and infinite systems with continuous spectra [42, 43]. Thus the charge-transfer reactivities defined by Parr and collaborators [8–12] exist and are useful only for such extended systems. The issue of how to define charge-transfer reactivities for the localized systems remains unresolved. We elaborate on this point below.

3. The Fukui function and the softnesses are charge-transfer reactivities. There are clearly chemical reactions in which no or little charge transfer occurs, for example the formation of homonuclear molecules or clusters or the formation of covalent entities with little ionic character. There is thus a need as well for isoelectronic reactivity indices which measure chemical response in the absence of charge transfer. How to define such reactivities emerges in Sect. 8 as a consequence of addressing and resolving issues 7, 8, and 9 for disjoint interacting reactants. This issue can therefore be considered resolved except for corrections to the reactivities of Sect. 8 which emerge from the overlap of the wave functions of the reactants. As discussed further below, the treatment of overlapping systems has not yet been carried out.

4. It was pointed out in [2, 3] that nuclear-configuration changes define chemical reactions so that nuclear reactivities should be defined and set on equal footing with the corresponding electronic reactivities. Thus nuclear Fukui functions ϕ_a, Eq. (59), and nuclear softnesses σ_a, Eq. (60), were defined in [2], and explicit Kohn-Sham expressions were found for them, Eqs. (61)–(64), as reviewed in Sect. 5. These are electron-transfer reactivities and are valid only for extended systems, leaving open the question of nuclear electron-transfer reactivities for localized systems and nuclear isoelectronic reactivities for all systems.

5. The issue of properly treating chemical response as a nonlocal response to a nonlocal perturbation was addressed in [3]. As discussed in Sect. 6 above, that work, through use of the Berkowitz-Parr softness kernel [15], Eqs. (66) and (67), resolved the issue for the charge-transfer reactivities of extended systems. The Nalewajski softness kernel [19–21], Eqs. (70) and (71a), on the other hand, resolves the issue for the isoelectronic reactivities of both localized and extended systems. For extended systems, the charge-transfer softness kernel $s^\mu(\mathbf{r}, \mathbf{r}')$ and the isoelectronic softness kernel $s^\mathcal{N}(\mathbf{r}, \mathbf{r}')$ are related via the Berkowitz-Parr relation, Eq. (72) [15]. An explicit Kohn-Sham expression for $s^\mu(\mathbf{r}, \mathbf{r}') = s^\mathcal{N}(\mathbf{r}, \mathbf{r}')$ for localized systems, Eq. (73), and $s^\mathcal{N}(\mathbf{r}, \mathbf{r}')$ for extended systems, Eq. (75a), are given here for the first time. The corresponding nuclear softness kernels, Eqs. (80)–(82), were introduced in [3] and reviewed in Sect. 6 as well.

6. As recognized by Nalewajski and Koniński [22], the softness kernels eliminate the ambiguities posed by the Fukui concept of frontier orbitals which persists in the Fukui function and the local and global softnesses. These kernels reflect the true complexity of chemical responses, as discussed in [3] and reviewed more briefly in Sect. 7 above.

7–9. By addressing issue 9 of how isoelectronic chemical reactivities enter the total energy of interaction of two disjoint, nonoverlapping systems [50], we show in Sect. 8 how contact can be made between chemical reactivities and the initial segment of, or portal to, the reaction pathway. As by-products, a clear definition of the chemical stimuli which couple to the reactivity indices, the chemical responses, and a clear definition of dynamic as well as static chemical reactivities emerge, resolving issues 7 and 8, respectively.

While all this represents substantial progress, several fundamental issues remain unresolved. Foremost among these, the resolution of which would lead to resolution of all others, is the definition of charge-transfer reactivities for localized systems, and possibly their redefinition for extended systems, which would emerge from generalizing the expression of the energy of interaction of two disjoint reactants, Eqs. (90), (91), (93), (94), and (96), to the case of overlapping reactants. Direct use of many-body perturbation theory led to these results, but this formalism becomes impracticably complex in the presence of overlap. Multiple scattering theory in principle offers a way out [53]. However, this theory is again impracticably complex in its many-body formulation. One is forced to go to the Kohn-Sham equations, which permit use of the independent-particle or single-particle formulation. The resulting problem would be straight-forwardly soluble were there precise knowledge of the Kohn-Sham potential $v_{KS}(\mathbf{r})$. Introducing both dispersion forces and overlap into the Kohn-Sham potential is, as is well known, an unsolved problem. We believe, however, that an astute combination of the methods of Sect. 8 with a multiple-scattering treatment of the Kohn-Sham equations could prove to be successful [53]. Were that to be the case, our goal of strengthening the foundations of reactivity theory as we presently understand it would have been reached.

Acknowledgement. I thank my colleagues M.V. Ganduglia-Pirovano and J. Kudrnovský for their contributions to the work reviewed here and V. Natoli for his assistance. I also thank M. Scheffler for discussions. I am especially indebted to R.G. Parr for stimulating comments and to R.F. Nalewajski for communicating and discussing his work.

10 References

1. Nalewajski RF, Korchowiec J, Michalak A (1996) Reactivity criteria in charge-sensitivity analysis. In: Nalewajski RF (ed) Topics in current chemistry (present volume). Springer, Berlin Heidelberg New York
2. Cohen MH, Ganduglia-Pirovano MV, Kudrnovský J (1994) J Chem Phys 101: 8988

3. Cohen MH, Ganduglia-Pirovano MV, Kudrnovský J (1995) J Chem Phys 103: 3543
4. Fukui K, Yonezawa T, Shingu H (1952) J Chem Phys 20: 722
5. Fukui K, Yonezawa T, Nagata C, Shingu H (1954) J Chem Phys 22: 1433
6. Fukui K (1975) Theory of orientation and stereoselection. Springer, Berlin Heidelberg New York
7. Fukui K (1982) Science 218: 747
8. Parr RG, Yang W (1989) Density functional theory of atoms and molecules. Oxford Univeristy Press, New York
9. Parr RG, Pearson RG (1983) J Am Chem Soc 105: 7512
10. Yang W, Parr RG, Pucci R (1984) J Chem Phys 81: 2862
11. Parr RG, Yang W (1984) J Am Chem Soc 106: 4049
12. Yang W, Parr RG (1985) Proc Natl Acad Sci USA 82: 6723
13. Ghosh SK, Berkowitz M (1985) J Chem Phys 83: 2976
14. Berkowitz M, Ghosh SK, Parr RG (1985) J Am Chem Soc 107: 6811
15. Berkowitz M, Parr RG (1988) J Chem Phys 88: 2554
16. Hohenberg P, Kohn W (1964) Phys Rev B 136: 864
17. Kohn W, Sham LJ (1965) Phys Rev A 140: 1153
18. Dreizler RM, Gross EKU (1990) Density functional theory. Springer, Berlin Heidelberg New York
19. Nalewajski RF (1984) J Chem Phys 78: 6112
20. Nalewajski RF (1984) J Chem Phys 81: 2088
21. Nalewajski RF (1985) J Chem Phys 89: 2831
22. Nalewajski RF, Koniński M (1987) Z Naturforsch 42a: 451
23. Nalewajski RF, Korchowiec J, Zhou Z (1988) Int J Quant Chem Symp 22: 349
24. Nalewajski RF (1993) In: Sen KD (ed) Structure and Bonding – Chemical hardness, Vol 80. Springer, Berlin Heidelberg New York, p 115
25. Miller WH (1994) In: Hänggi P, Talkner P (eds) New trends in reaction rate theory. Kluwer, Norwood, MA
26. Cohen MH, Natoli V, Ganduglia-Pirovano MV, Kudrnovský J (1995) Chemical reactivity for physicists; a work in progress. In: Chelikowsky JR, Louie SC (eds) Quantum theory of real materials. Kluwer, Norwood, MA
27. White JA, Bird DM (1993) Chem Phys Lett 213: 3
28. White JA, Bird DM, Payne MC, Stich I (1994) Phys Rev Lett 73: 1404
29. White JA, Bird DM, Payne MC (1995) Phys Rev B (submitted)
30. Hammer B, Jacobsen KW, Nørskov JK (1992) Phys Rev Lett 69: 1971
31. Hammer B, Scheffler M, Jacobsen KW, Nørskov JK (1994) Phys Rev Lett 73: 1400
32. Hammer B, Scheffler M (1995) Phys Rev Lett 74: 3487
33. Wilke S, Scheffler M (1995) Phys Rev B (submitted)
34. Wilke S, Scheffler M (1995) (in preparation)
35. Gross A, Hammer B, Scheffler M, Brenig W (1994) Phys Rev Lett 73: 3121
36. Gross A, Wilke S, Scheffler M (1995) Phys Rev Lett (in press)
37. Equation (3) differs slightly from the expression for $f^-(r)$ in [8–12], which contains a minor error corrected in [2]
38. For consistency and precision we have added the superscript $\pm$ omitted in [8–12]
39. Here the distinction between $+$ and $-$ must still be kept because a general variation $\delta\rho(r)$ in $\rho(r)$ implies a variation $\delta\mathcal{N}$ in $\mathcal{N}$, $\int dr \delta\rho(r) = \delta\mathcal{N}$
40. Yang and Parr were the first to point to a relation between softness and density of states in [12], but they, in effect, replaced the kernel $\mathcal{K}^{-1}(r, r'')$ Eq. (21) by $\delta(r, r')$
41. Note the distinction made here between the KS potential in Eq. (17) and the KS interaction in Eqs. (23) and (26) and similarly for the exchange-correlation potential, Eq. (19), vs. the exchange-correlation interaction Eqs. (26b, c).
42. A more detailed report of the material presented in Sects. 4.1–4.4 will be published elsewhere.
43. The subjects of Sects. 4.1–4.3 have also been discussed by Gazquez JL, Galvan M, Vela A (1990) Theor Chem 69: 29
44. Cohen MH, Natoli V (1995) presented at 19th Int Conf on Statistical Physics, Xiamen, China and to be published [42]
45. Nakatsuji H (1973) J Am Chem Soc 95: 345
46. Nakatsuji H (1974) J Am Chem Soc 96: 24
47. Nakatsuji H (1974) J Am Chem Soc 96: 30

48. Feynman RP (1939) Phys Rev 56: 340
49. Zimmerman HE, Weinhold F (1994) J Am Chem Soc 116: 1579
50. Cohen MH, Ganduglia-Pirovano MV (1996) (in preparation)
51. Hirschfelder JO, Meath WJ (1968) Adv Chem Phys 12: 3
52. Mahan GD, Subbaswamy KR (1990) Local density theory of polarizability. Plenum, New York p 109ff
53. Cohen MH, Ganduglia-Pirovano MV (unpublished)

Author Index Volumes 151–183

The volume numbers are printed in italics

Springer-Verlag
and the Environment

We at Springer-Verlag firmly believe that an international science publisher has a special obligation to the environment, and our corporate policies consistently reflect this conviction.

We also expect our business partners – paper mills, printers, packaging manufacturers, etc. – to commit themselves to using environmentally friendly materials and production processes.

The paper in this book is made from low- or no-chlorine pulp and is acid free, in conformance with international standards for paper permanency.